A Hand Papermaker's Sourcebook

Sophie Dawson and Silvie Turner

Acknowledgements

Sophie Dawson and Silvie Turner would like to thank everyone around the world who supplied information and photographs, gave permissions, advice and helped make this book possible. estamp would like to thank Simon Green for so thoroughly proof-reading the text.

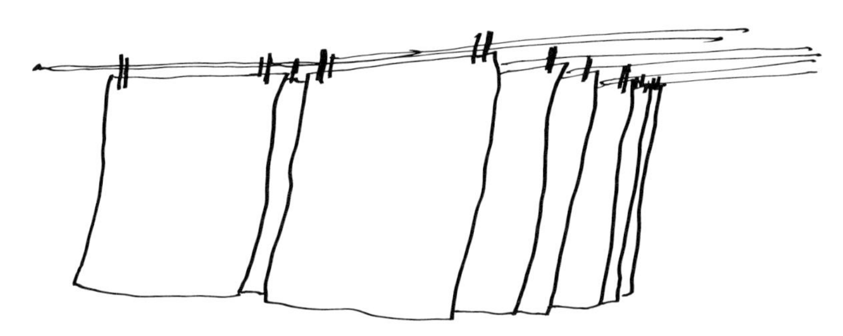

Designed by Lone Morton. Printed by BAS Printers Ltd., Over Wallop, Hampshire SO20 8JD, England

British edition
First published in Britain 1995
A catalogue record for this book is available from the British Library
Published by estamp 204 St Albans Ave, London W4 5JU
Distributed in UK by Central Books 99 Wallis Road, London E9 5LN
Telephone 0181 986 4854 Fax 0181 533 5821
ISBN 1 871831 14 8

American edition
First published in the United States 1995
Library of Congress Cataloguing-in-Publication Data
Dawson, Sophie
A hand papermaker's sourcebook / Sophie Dawson and Silvie Turner.
p. cm.
Includes index.
ISBN 1-871831-14-8 (Estamp). - - ISBN 1-55821-389-9 (Design Books)
1. Papermaking. 2. Paper, Handmade. I. Title.
TS1105.D313 1995.
676'.22--dc20 95-14862
CIP

Published by Design Books
Design Books are distributed in the USA and Canada by Lyons & Burford, Publishers
31 West 21 Street, New York, NY 10010

Contents

Introduction

Kathryn Clark of Twinrocker, USA

. . . no wonder papermaking was held in such awe in bygone days and was known as the 'white craft', for it is a kind of magic - even a kind of miracle . . .

Jean Chitty *Paper in Devon* 1976

The making of paper by hand is burgeoning. Aspects of papermaking range from custom making - where craftsmen making sheets are known to concentrate on subtlety, archival soundness, evenly-made paper quality and to the value the nature of the craft and process - to artists who more frequently focus on immediacy, variability, and the pushing of boundaries, when tactile and visual properties take precedence often with a disregard for established methods and ideas of permanence. With the demand for papermaking teachers and workshops and the expansion of awareness of handmade paper, it seems an opportune moment for well-crafted equipment and materials to become more widely available. What to buy, where to find it, where to learn the processes, where to make contact with others - we have identified a network of international suppliers around the world and carefully compiled the product descriptions in accordance with the information given to us. Many of the larger companies we list offer a comprehensive range of papermaking goods and services which are thoroughly tested, and also provide information to help you choose and use these materials. Most manufacturers and suppliers are mail order-based, produce detailed catalogues and welcome specific enquiries - some even give free technical advice over the telephone. Smaller manufacturers and distributors tend to specialise in particular items, for example, papermaking kits, moulds and deckles, beaters and pulp-processing equipment. The currency of information in a book such as this is always at risk because companies constantly expand and decline and we apologise in advance for any errors. The listings, we hope, will not only give papermakers a source of supply but will also act as a reference to the many diverse aspects of hand papermaking that have developed during the past few years. We are conscious that there are gaps. Language has presented a barrier to our collection - we are sure that papermakers, like us, may often find that communication is not always a verbal facility! However, estamp would like to hear from papermaker's suppliers in other countries that have not been covered by this first edition and we expect that both readers and suppliers will keep us informed with details for future editions. We hope you will like the idea of this book. Although it is a directory of information, it is also much more. To us, it's a picture of a contemporary craft. We have discovered that often the best products have been found, invented, developed or designed by the papermakers themselves or their immediate family and friends. If one looks back into the past, it would be called a 'trade'. For us it seems very satisfying that this 'trade' continues in a contemporary context fulfiling one of man's inherent needs - to create, to make and to supply.

Sophie Dawson and Silvie Turner

Telephone dialling

We have listed telephone numbers given to us by suppliers. These numbers include the area code (in brackets) and the suppliers' numbers. If you wish to dial a number in another country, the codes below will help you. The same rules apply for faxes.
INTERNATIONAL DIRECT DIALLING is based on a standard system comprising of four elements :
FIRSTLY The international code. See below.
SECONDLY The country code. See below.
For example, to use our list - if you are dialling from UK to USA, use the international code from UK = 00, followed by national code for USA = 1.
THIRDLY The area code. Note - When dialling an international number, if an area code begins with an '0' (e.g. 0171), it is necessary to leave the '0' out of the code.
FOURTHLY The suppliers' number, given by us.

INTERNATIONAL DIALLING CODES

	Dialling from (int.code)	*Dialling to (country)*
Australia	0011	61
Austria	00	43
Belgium	00	32
Canada	011	1
Denmark	009	45
Egypt	00	20
Finland	990	358
France	19	33
Germany	00	49
Greece	00	30
Ireland	00	353
Israel	00	972
Italy	00	39
India	00	91
Japan	001	81
Netherlands	00	31
New Zealand	00	64
Norway	095	47
Portugal	00	351
South Africa	09	27
Spain	07	34
Sweden	009	46
Switzerland	00	41
UK	00	44
USA	011	1

Measurements

We have listed feet, inches, millimetres, centimetres, square feet, etc. reflecting suppliers' information.

METRIC CONVERSION FACTORS
1 millimetre (mm) = 1000 micrometres = 0.0394 inch (in)
1 centimetre (cm) = 10 millimetres = 0.3937 inch
1 metre (m) = 100 centimetres = 1.0936 yards (yd)
1 inch = 2.54 centimetres
1 foot (ft) = 12 inches = 303.48 centimetres
1 yard (yd) = 36 inches = 0.9144 metres
1 square metre (sq.m) = 10,000 square centimetres (sq.cm) = 1.196 square yards (sq.yd)

TO CONVERT	MULTIPLY BY
inches to centimetres	2.54
centimetres to inches	0.3937
feet to metres	0.3048
metres to feet	3.2808
yards to metres	0.19144
metres to yards	1.09361

1 gram (gm) = 1000 milligram (mgm) = 0.0353 ounce (oz)
1 kilogram (kg) = 1000 gram = 2.2046 pounds (lb)
1 ounce = 437.5 grains(gs) = 28.35 grams
1 pound = 16 ounces = 0.4536 kilogram
1 pint (pt) = 0.8327 UK pint = 0.4732 litre (ltr)
8 pints = 1 gallon (gal) = 3.778 litres

INTERNATIONAL PAPER SIZES
AO 841 x 1189 mm = 33.11 x 46.81 in
A1 594 x 841 = 23.39 x 33.11
A2 420 x 594 = 16.54 x 23.39
A3 297 x 420 = 11.69 x 16.54
A4 210 x 297 = 8.27 x 11.69
A5 148 x 210 = 5.83 x 8.27
A6 105 x 148 = 4.13 x 5.83

INTERNATIONAL ENVELOPE SIZES
C4 229 x 324 mm = 9 x 12.75 in
B4 250 x 353 = 9.88 x 12.88
C5 162 x 229 = 6.38 x 9
B5 176 x 250 = 7 x 9.88
C6 114 x 162 = 4.5 x 6.38
B6 125 x 176 = 5 x 7

. . . at all costs this hand craft must be preserved.

J Barcham Green *Papermaking by Hand* in 1953 (Out of print)

. . . the manufacture of paper by
hand only requires dextrous love
and strong sacrospinalis . . .

Walter Hamady in Jules Heller *Papermaking* 1978

Making paper
Studio needs
General studio supplies

Making paper

Paper is made from cellulose found in plants and trees. Cellulose is the principal component of the cell walls of most plant life, which we call the fibre. Before they can be used to make paper, the fibres must be liberated from the plant and any non-cellulose material must be removed. This is usually done by cooking the fibres in an alkaline solution with, occasionally, a pre-cooking stage called fermenting (or retting).

After cooking, the next step in the preparation process is to beat the fibres with water into a pulp. Beating softens the fibres, unravels the smaller fibre components called fibrils, and works water into their intermolecular structure. This makes possible the chemical bonding between the fibres (known as hydrogen bonding) that holds a sheet of paper together.

When the fibre has been sufficiently beaten, it is diluted with water in a vat. The standard European method of forming a sheet of paper is by dipping a mould and deckle, a wooden frame with a wire screen or mesh attached, into the vat mixture. As the mould is lifted from the vat, the water begins to drain and the papermaker shakes the mould

it is possible to know the mill by looking at its papers . .

G. A. Beale *A Survey of Hand-made and Mouldmade Papers* 1977 (Out of prir

slightly, causing the fibres to interlock in all directions, in an even layer on the mould surface. Once the water has drained, the deckle is removed, and the sheet of paper is transferred from the mould in a process called 'couching'. When a pile of freshly couched sheets, a 'post', is completed, it is loaded into a press. The press creates a physical bond between the fibres and begins the drying process by expelling as much water as possible from the sheets of paper.

Successful drying is best achieved slowly to allow the fibres to shrink naturally and secure the bonds which have already been formed in the papermaking step. This, in turn, gives the finished sheet greater dimensional stability and a good folding strength.

Once dried, the sheets are sized, or made water-resistant, by one of two methods. The first, traditional approach, involves immersing a stack of dried sheets into a prepared solution of gelatine, after which they are pressed and dried a second time. This method is known as 'tub' or external sizing. The second method involves adding an internal sizing agent to the pulp in the last few minutes of the beating cycle.

Until the beginning of the nineteenth century, all paper was handmade in sheet form. Although the ancient traditions of the hand craft have been transformed into a modern industry, the best fine papers are still those where the careful consideration of particular characteristics, dictated by the specific use to which the sheet is to be put, remain a primary concern.

Studio needs

Papermakers can work at home on their kitchen tables, in bedrooms or basements, in classrooms, in sheds, out in back gardens or in specially-constructed studios. Advice to papermakers about what equipment is needed for papermaking varies widely depending on whom you ask. From a papermill in Huxam, Exeter, it was recorded in 1703 that the requisites for a one-vat paper mill were: 2 presses, 1 furnace, 1 large cistern, 3 chests, 48 trebbles with lines, 1 great water wheel, 1 cog wheel, 1 x five-hammer mortar, 4 x four-hammer mortars, 1 pit mortar, 9 pairs of moulds with 2 deckles each, 7 pairs of moulds with 3 different deckles, felts, basins, strainers.

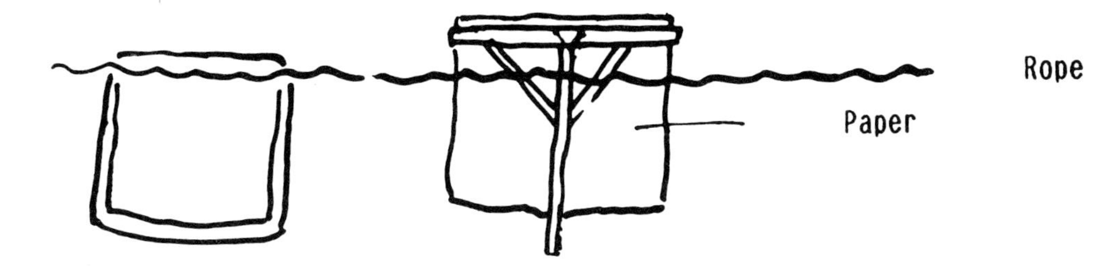

A basic papermaking studio today will use any of the following :
adhesives, apprentice moulds, basic moulds and deckles, a beater, bleaches, books, casting equipment, chemicals, cooking pots, a dehumidifier, a drying machine, drying areas, dyes, fibres, felts, hot plates, mixers, pigments, presses, ready-to-use pulps, a sink, safety equipment, size, a vacuum system, vats and water. Additional miscellaneous equipment such as buckets, stainless steel or enamel pots, cloths, mops, measuring bowls, irons, palette knives, clamps, pressing boards, trays, spoons, scissors, sticks, knives, wire, waterproof boots, rubber aprons, etc. can all be bought from your local hardware, kitchen, garden or even butcher's supply shops.

Sophie Dawson making paper

. . . water has been vital to man and nature since the beginning of time . . . water is never pure . . . and its impurities are the factors of concern.

Drew *Principles of Industrial Water Treatment* 1977

Water has always been one of the crucial factors in papermaking. As water goes through its cycles it picks up impurities which reflect the composition of the atmosphere and the earth's crust. Rainwater can dissolve gases such as oxygen, carbon dioxide (non-industrial), sulphur and nitrogen dioxide (industrial) as well as absorb soluble and insoluble matter. It may also contain organic matter caused by industrial contamination of the atmosphere as well as that evolved from vegetation and even contain concentrated deposits of limestone, magnesite, iron ore, gypsum, copper ore, sulphur and other compounds.

General studio supplies

AUSTRALIA

■ THE PAPER MERCHANT
316 Rokeby Road
Subiaco, WA 6008
Tel: (09) 381 6489
Fax: (09) 381 3293

Many miscellaneous supplies. Write for catalogue.

CANADA

■ KAKALI HANDMADE PAPERS, INC.
1249 Cartwright Street
Vancouver, BC V6H 3R7
Tel: (604) 682 5274

Bone Folders

Buckets with locking lids (2.5gal capacity)

8oz Squeeze Bottles

■ THE PAPERTRAIL
Handmade Paper & Book Arts
1546 Chatelain Avenue
Ottawa, Ontario K1Z 8B5
Tel: (613) 728 4669
Fax: (613) 728 7796
Orders: 1 800 363 9735

Includes everything for the papermaker including strainers, bottles, aprons, gloves, bone folders, stainless steel strainers, squeegees and barrier creams. Also squeegees with curved ends (handles not included). They communicate in English, French, German and Spanish.

UK

■ MERCK LTD.
Head office Broom Road
Poole, Dorset
Tel: (01202) 745520

pH Test Strips
A paper for measuring the pH values.

■ TEST PAPERS DIVISION
The Paterson Photax Group
The Gate Studios
Station Road
Boreham Wood
Herts WD6 1DQ

pH Test Strips
Universal Papers for measuring pH values.

USA

■ CARRIAGE HOUSE PAPER
79 Guernsey Street
Brooklyn, New York 11222
Tel/Fax: (718) 599 PULP (7857)
Orders: 1 800 669 8781

Bone Folders

Chiri-tori Tweezers
Stainless steel forceps perfect for picking out black bark from Japanese fibres. Invaluable for extracting pulp from the intricacies of moulds and the crevices of a Hollander beater or separating layers of wet pulp in a laminated art work.

Plastic Squeeze Bottles (8oz capacity)

Pulp Painters
Syringe with a curved tip, which can be cut to any size opening.

Pump Dispensers
A simple, mess-free way of dispensing liquids such as PVA glue and retention agents.

Stainless Steel Strainers
Industrial-type, stainless steel, conical strainer, measuring 11.5in across the top, with a hook at one edge and a 9in handle, so that the strainer can rest securely on top of a bucket.

Stainless Steel Cooking Pots
With lids and handles for cooking fibres with soda ash or lye.

Thomas Water Filtration System
Attached to the end of a hose to provide filtered water to a vat or beater.

■ GOLD'S ARTWORKS, INC.
2100 North Pine Street
Lumberton, NC 28358
Tel: (919) 739 9605
Orders: 1 800 356 2306

Absorb'n Dry PVA Block
Size 6.5 x 2.75in; absorbs water without dripping; for blotting cast pieces/all forms of paper.

5gal Buckets

Mortar and Pestle Sets

pH Test Strips

Portable Hand Wringers
Rustproof, all-steel frame, steel-tube handle and zinc-plated finish. Hard maple bearings never needing oiling. A single adjustable screw applies pressure at the centre of a flat, tempered-steel spring so pressure is balanced over the entire length of the roll.

Synthetic Chammy
A spun rayon fabric sold as an all-purpose cloth for wiping up spills. Absorbs almost twice its own weight of water.

■ LEE S. McDONALD, INC.
PO Box 264
Charlestown, MA 02129
Tel: (617) 242 2505
Fax: (617) 242 8825

Flexo-Felt Brushes made of Palmyra bristles and used to keep felts clean and free of pulp; pH Test Strips, Pulp Painting Bottles. Also Naga Brushes for brushing newly made paper onto boards, Hand Beating Sticks for hand processing bast fibres; Felt Brushes; Marbling supplies and books on papermaking, etc.

Pyro-Pruf Flame Retardant
A liquid flame retardant with a pH of 7.4 to 7.6 for paper and cellulose fabrics, Pyro-Pruf can be brushed or sprayed onto a dry surface to reduce flammability. It is completely water-soluble.

■ MAGNOLIA EDITIONS
2527 Magnolia Street
Oakland, CA 02129
Tel: (510) 839 5268
Fax: (510) 893 8334

Elastic Top Strainer Bags
A fine-weave-fibre gallon nylon bag recommended for straining coagulant and formation aid to prevent undissolved particles from contaminating the vat.

Plastic Trays
Perforated trays used as a stencil for making lace paper or as a drain for rinsing pulp.

. . . whoever thinks of going to bed before twelve o'clock is a scoundrel. Samuel Johnson

Pump Sprayers
A versatile and durable pressurised sprayer.

Strainers/Colanders
1 Stainless Steel Sieve. Ideal for straining pulp or making circular paper 11in diameter; three removable screens: fine, medium and coarse weave.
2 Plastic Colander. Perfect for straining and rinsing most types of pulp.

Also Drying Brushes, 5gal Buckets, Plastic Pulp Pots, Plastic Squeeze Bottles in various sizes, Sponges and Tweezers.

■ PRO CHEMICAL AND DYE INC.
PO Box 14
Somerset, MA 02726
Tel: (508) 676 3838
Fax: (508) 676 3980

Beakers
Polypropylene plastic TRI-POUR safety beakers. Measuring pipettes for accurate measuring of stock solutions.

pH meters
Electronic, digital pH meter saves the cost of pH paper.

Plastic Squeeze Bottles
Polyethylene bottles in 4 convenient sizes with closed tips that can be snipped to a desired opening size.

Fibres and pulps

Left top Helmut Becker hand pulling flax fibre
Left bottom The surface of a sheet of paper at magnification x 1200
Right top Magnification of linen fibre
Right bottom Hollander beater containing cotton linter pulp

Fibres and pulps

Fibres and pulp for papermaking are available in a number of different forms. We have listed two distinct categories of fibres supplies: firstly raw unprocessed fibres and secondly partly processed fibres and ready-to-use pulps.

. . . any plant that grows is certainly eligible - the only qualifying condition is the quantity of cellulose fiber in relation to the quality of paper. . .

Walter Hamady in Jules Heller *Papermaking* 1978

RAW UNPROCESSED FIBRES

Raw fibres are usually in the form of long, loose strands and are classified according to their location in the plant. They require some kind of treatment, usually cooking in an alkaline solution to remove non-cellulose impurities as well as beating to render them suitable for papermaking. Bast or inner bark fibres, such as flax, hemp and *kozo*, contain some of the longest fibres for making paper. Leaf stem fibres, such as abaca, and leaf fibres from plants such as sisal and yucca, offer a comparatively shorter range of fibre but still produce a strong, crisp paper. A further category of fibres are attached to the covering around the seeds of certain plants. Kapok and cotton, for example, are familiar examples of 'seed-hair' fibres.

PROCESSED FIBRES

Processed fibres are cooked and partially beaten fibres (which we call 'half-stuff') that are sold in dry, compressed sheet form and as woven rags. Cotton and linen fabric scraps, or rags from the textile industry, are a useful source of fibre for making paper. Rag fibre, made from new garment cuttings, is also called 'half-stuff' because the fibres have been broken out of the cloth into near thread form and are therefore half-prepared.

READY-TO-USE PULPS

These are sometimes pre-beaten to customer specification and are a convenient source if you do not have access to beating and/or processing equipment.

Fibres

ABACA
(*Musa textilis*)

Abaca is the Philippine word for Manila hemp. This fibre comes from the stalk of a plant related to bananas. It has a long fibre length and is a versatile, all-purpose pulp that can be used for making Western paper as well as thin sheets of Oriental-style paper. In sheet form it is easily rehydrated in a blender; but it may also be beaten in a Hollander to produce a harder, stronger paper. It is excellent for pulp painting or pulp spraying and can be blended with shorter fibres to impart additional strength.

COIR
(*Cocos nucifera*)

Coir is the fibre from coconut husks traditionally used to make doormats. It is usually available in long, loose strands and must be cooked before beating to be used for papermaking. It is an unusual brown colour and can be mixed with other fibres to provide an interesting fleck.

COTTON
(Gossypium sp.)

Cotton is a strong, versatile, 'seed-hair' fibre that comes in many forms. When the cotton is ginned, the long 'staple' fibres (raw cotton) are separated from the seeds and used to make cotton cloth.

COTTON RAG PULP

Cotton rag pulp is made from cotton cloth (already spun and woven)which has usually been cooked, shredded and partially beaten. Some suppliers sell different types of clothing, already beaten, e.g. 'blue jeans'.

COTTON LINTERS

The shorter fibres which cover the remaining seeds after the staple cotton has been removed are the source for cotton linters. Several types of linters are available: 1st Cut is recommended for sheet forming and is also suitable for casting as it shrinks little in drying. 2nd Cut is a shorter fibre that is used for sheet forming but is also useful in picking up details in sculptural papermaking. Mill-run Grade is a mixture of 1st and 2nd Cut. It processes more easily than a pure 1st Cut and is a good choice for better quality papers that require a medium fibre length.

UNBLEACHED COTTON FIBRE

A new type of cotton is raw cotton, not linters. This is not processed in any way (except ginning) to remove the seeds. It has been grown in the US by FoxFibre®, a new company that develops fibres for the textile industry.

ESPARTO
(Lygeum spartum)

Esparto is a fibre from the esparto grass plant. The fibre has historically been used in the manufacture of rope, shoes (espadrilles received their name from the fibre) and baskets. A relatively short fibre which is not terribly strong, but because it shrinks very little in drying it is a good casting fibre.

FLAX
(Linum usitatissimum)

Flax is a herbaceous bast fibre and is used to produce linen cloth. Three types are generally available: raw (uncooked) flax, flax half-stuff and linen rag.

Raw flax is divided into two grades: Line Flax, which is combed clean of any straw and used for linen cloth, and Flax Tow. When retted, the flax is scutched and produces 'line flax' which after combing, roving and spinning becomes linen thread. The shorter fibre (anything less than full-length fibre) is called 'tow' and is processed as a separate industry. Most straw is removed by scutching and combing, however it is often accepted that 'there is no flax without shirle'. Raw flax must be cooked before it can be used for papermaking. It can be used to make incredibly strong, translucent paper and is excellent for pulp painting and pulp spraying. With very long beating times, flax and linen exhibit a high degree of shrinkage in drying, a property that can be utilised when creating three-dimensional forms.

Flax half-stuff is usually a European flax fibre that has been cooked and partially processed in a beater. It comes in sheet form, ready for further preparation.

Linen rag comes from a quality linen, spun, woven and from combed line flax.

HEMP
(Cannabis sativa)

Hemp is native to China and one of the longest and strongest fibres for papermaking. It is thought to be the original fibre used to make paper in China two thousand years ago.

KAPOK
(Ceiba pentandra)

Kapok is a hollow, 'seed-hair' fibre that grows in a pod on a tree found in the South Pacific. It is a silky, cream-coloured fibre that can be used to make translucent paper when formed into a thin sheet. It is usually available in long, loose strands, ready for cooking and beating. Because the fibre is hollow and contains pockets of air, it must be crushed before it is cooked to prevent it from floating in the water.

KENAF
(Hibiscus cannabinus)

Kenaf is a bast fibre. It produces a very fine, strong beige-coloured paper. Kenaf needs to be cooked in lye and beaten in a Hollander beater for papermaking.

RAMIE

Ramie is a bast fibre from the stem of a relative of the nettle family.

SISAL
(Agave rigida)

Sisal is a leaf fibre that has traditionally been used for rope-making. Its long fibres produce a slightly softer and whiter paper than abaca.

WHEAT STRAW
(Triticum spp.)

Wheat is the most important of all cereal grasses and is cultivated. Straw pulp is produced after cooking with caustic soda. Its value as a papermaking fibre varies with the cellulose.

NOTE For Japanese, Thai and Philippine bark fibres see ORIENTAL PAPERMAKING.

NOTE Trade textile sources and rope works (not listed here) may yield positive results in the search for fibres for papermaking, for example, a source of raw abaca comes from unbleached tea bags but suppliers will probably only sell in large quantities so a 'consortium' may be needed to buy from larger suppliers. Remember also hand weavers' studios as a possibility for inclusions.

NOTE Local hand papermaking studios, workshops and mills will often be a source of small quantities of fibres and pulp but they are not listed here.

NOTE Many countries have a government-funded help-service to aid a small firm's search for suppliers in other small business areas, such as the Small Business Helpline in UK (contact Small Firm Division, Department of Employment).

Raw fibres

ARGENTINA

■ EL MOLINO DEL MANZANO
Fondo de la Legua 375
1642 San Isidro
Buenos Aires
Tel: (1) 763 8682

Flax

New Zealand Flax (*Phormium tenax*)

AUSTRALIA

■ PAPER CAPERS
PO Box 281
Neutral Bay, NSW 2089
Tel: (02) 964 9471

Indigenous fibres including New Zealand Flax and Banana. Contact De-Arne King.

CANADA

■ CRANBERRY MILLS
R.R. No.1
Seeleys Bay
Ontario K0H 2NO
Tel: (613) 387 1021

Cotton Linters
No2 pulp. Small (0.5 to 10kg) quantities.

Also pulp and paper testing facilities available. Contact the proprietor, Ted Snider.

■ KAKALI HANDMADE PAPERS, INC.
1249 Cartwright Street
Vancouver, B.C. V6H 3R7
Tel: (604) 682 5274

Unbleached Flax Fibre
A high-quality flax in raw form. Uncooked beaten fibre results in a crisp paper, and cooked gives a softer and lighter-colour paper.

■ MIKOLET STUDIOS
RR #44
Hampton, NB EOG 1ZO
Tel: (506) 832 3868

New, fully-equipped papermil Some fibre /pulp sales. Contact Colette Johnson-Vosberg for details.

■ THE PAPERTRAIL
1546 Chatelain Avenue
Ottawa, Ont. K1Z 8B5
Tel: (613) 728 4669
Fax: (613) 728 7796
Orders: 1 800 363 9735

Abaca
Long strands of fibre which must be cooked and beaten in a Hollander beater.

Unbleached Flax Noils
Loose fibres which must be cooked and beaten in a Hollander beater. Natural greyish-brown colour.

Flax Tow; Jute; Sisal. (All come in long strands which must be cut, cooked and beaten in a Hollander beater.)

■ HEMPLINE INC. 10 Gibson Drive Tillsonburg, Ontario N4G 5G5 Tel: (519) 842 9344	Information on hemp growing, processing and fibre; send a SSAE to Geof Kime, Director.

EGYPT

■ ASSOCIATION FOR THE PROTECTION OF THE ENVIRONMENT Zabbeleen Paper Centre Marie Assaad/ Isis Bailey Manshiet Nasser, Mokattam PO Box 32 Qalaa Cairo Tel: (202) 354 3305 Fax: (202) 273 5139	Banana (*Musaceae*); Common Reed (*Phragmotes australis*); Water Hyacinth (*Eichornia crassipes*)

GERMANY

■ EIFELTOR MÜHLE Auf dem Essig 3 D-53359 Rheinbach-Hilberath Tel: (22 26) 2102 Fax: (22 26) 2102	Contact Claudia Stroh Gerard. Eifeltor Mühle is a mail order service for hand papermakers and paper artists. They offer all handmade papermaking supplies including cotton linters (bleached and unbleached), hemp, flax, abaca. Contact mill for more details.

SWEDEN

■ ZEN ART PAPER AB Christina Bolling St Jorgens vag 20 S-422 49 Hisings Backa Tel: (31) 55 68 55 Fax (31) 55 56 86	Banana ('Torrade banana leaf'); Nepalese Daphne (Lokta bark) fibre; Jute

UK

■ BARCHAM GREEN & COMPANY LIMITED Hayle Mill Maidstone, Kent ME15 6XQ Tel: (01622) 692266 Fax: (01622) 756381	Flax Bleached and unbleached. Black Cotton Rags

■ KHADI PAPERS
Unit 3, Chilgrove Farm
Chilgrove, Chichester
PO18 9HU
Tel: (01243) 59314
Fax: (01243) 59354

Nepalese Daphne (*Lokta* bark) Fibre
Contact Nigel Macfarlane.

■ FRANK MONKMAN LTD.
Marshfield Mills
Marsh Street
Bradford BD5 9NQ
Tel: (01274) 723794

Flax Noils
1 Optic bleached, 2 Non-optic.

USA

■ CARRIAGE HOUSE PAPER
79 Guernsey Street
Brooklyn, New York 11222
Tel/Fax: (718) 599 PULP (7857)
Orders: 1 800 669 8781

Bark Fibres
Japanese *Kozo*; Thai *Kozo*; *Mitsumata*; Japanese *Gampi*; Phillipine *Gampi*.
(See ORIENTAL PAPERMAKING)

Raw Cotton Fibre
Organically grown, naturally coloured raw cotton (not linters).

Unbleached Flax Fibre
A high-quality flax fibre which has been cut into 0.25in lengths, making it less likely to tangle in a beater. The flax may be beaten without cooking, resulting in a crisp, rattly paper; or it may be cooked to produce a softer sheet, lighter in colour.

Also Hemp and Kenaf Fibre

■ GOLD'S ARTWORKS, INC.
2100 North Pine Street
Lumberton, NC 28358
Tel: (919) 7399605

Raffia
Dry grass fibre (requires cooking).

Sea Grass
Twisted bundle (requires cooking).

■ LEE S. McDONALD, INC.
PO Box 264
Charlestown, MA 02129
Tel: (617) 242 2505
Fax: (617) 242 8825

Abaca Fibre
Two types available: 1 Philippine Abaca, 2 Ecuadoran Abaca.

Bark Fibres
Japanese *Kozo*; Thai *Kozo*; *Mitsumata*; Japanese *Gampi*; Philippine *Gampi*.

Flax
Available in the following forms: 1 Flax sliver, 2 Flax noils, 3 Flax tow; various grades, 4 Chopped Flax.

True Hemp
In the following forms: 1 Hemp fibre, 2 Hemp tow, 3 Retted Hemp, 4 Green Hemp.

Ramie
Available in two forms: 1 Sliver, 2 Other various grades.

Special Order Fibres include Bagasse; Bogang; Coir; Esparto; 'Sunn Hemp'; Jute; Maguey; Salago. Please contact supplier for more detail about these fibres.

■ MAGNOLIA EDITIONS
2527 Magnolia Street
Oakland, CA 94607
Tel: (510) 839 5268
Fax: (510) 893 8334

Bark Fibres
Thai *Kozo*; Korean *Kozo*; Philippine *Gampi*; Japanese *Mitsumata*. (See ORIENTAL PAPERMAKING)

■ MARALEX STUDIOS
Graybridge
3251 Fernwood Street
Arden Hills, MN 55112

Indigenous fibres from Jamaica including Banana; Sisal; Wildcane; Sansevieria; Bamboo; Bagasse and Almond fibres; also Minnesota Cattail Fibre.

■ STRAW INTO GOLD
3006 San Pablo Avenue
Berkeley, CA 94702
Tel: (510) 548 5247
Fax: (510) 548 3543

Combed Flax Top (bleached white)
Soft, lustrous and extra fine.

Combed Flax Top (natural)
Top-quality, dew-retted flax from France. Cut into long staple-length, carded and combed with a lustrous tan colour.

Flax Strick (natural in hank)
Also top-quality French flax which has been hackled, but not cut from its original length (about 18in to 24in).

Ramie (combed top)
Creamy, natural-white colour.

■ TWINROCKER HANDMADE PAPER
PO Box 413
Brookston, IN 47923
Tel: (317) 563 3119
Fax: (317) 563 8946

Twinrocker specialises in fibres and pulps and also offers pigments and sizing.
Raw fibres include:
Coir
Long, loose strands of raw fibre ready for cooking and beating.

Raw Cotton
Loose strands of unbleached fibre ready for cooking and beating.

Raw Line Flax
This pure line flax has been cut to 2in lengths for easier cooking and beating. It is a natural tan colour.

Raw Flax Tow
This is an unrefined fibre with straw and other impurities. Long, loose strands of fibre ready for cooking and beating.

Russian Hemp

Kapok
Available in two forms - 1 Pale with specks, 2 Dark with more specks.

Sisal
Long golden strands of fibre, ready for cooking and beating.

Wheat Straw
Golden wheat straw from the Brookston area of Indiana. It makes non-archival paper. Cooking removes impurities but leaves most of the colour.

Plant fibres

Brian and Maureen Richardson operating the chaff cutter, cutting rushes into small pieces

We have no section on the availability of fibres from plants because local plant sources are available to everyone. There are papermakers and mills which specialise in producing papers from plants and we have taken the following information from *Plant Papers*, the name of a small booklet (and incidentally also the name of her mill) by Maureen Richardson who works exclusively with plant fibres in Herefordshire, England. Although Maureen does not supply fibres, she runs workshops on making papers from plant sources. "Almost any plant will do for papermaking. Plants which make particularly good pulps are straw, nettle and rush. Others also suitable include pampas grass, the long leaves of the iris, red hot poker, monbretia and gone to seed annuals, vegetables and tomato haulms. Equipment necessary includes a galvanised or stainless steel bucket, nylon sieve, wooden spoon, rubber gloves, old scissors, mallet or hammer, carbonate of soda (washing soda for softer fibres, e.g. flowers) or caustic soda (sodium hydroxide for tougher fibres), gas or electric ring and containers with lids." PLANT PAPERS Romilly, Brilley, Herefordshire.

Prepared pulps

This section includes pulps that are prepared in some way and includes 'half-stuff' and fully prepared pulps. Simon Green quite rightly points out that we have a dilemma with the term 'half-stuff'. He cites Labarre's traditional definition:
"Half-stuff, formerly also known as first stuff, is the term applied to any partially broken and washed stock (rag pulp or other material) and thus reduced to a fibrous pulp, usually before it is bleached. Half-stuff is a wet, in-mill material and was not something usually traded or dried although it could be kept in the mill drained off, or as press pâte until it became revolting!"
As the pulp suppliers use a variety of terms, we have decided to keep this section a general one on the availability of fully- or semi-processed pulps. We strongly suggest that when ordering prepared pulps you check with your suppliers that you are getting what you want, e.g. exactly the right type of pulp, the beating time that has taken place, the amount, sizing (or not) that has been added, pigment or colour addition (also any tax, packing costs, shipping costs or any other hidden costs).

NOTE There are different types of pulps, for example, for sheet forming, casting pulp, painting pulp, pulp spraying, etc., and pulps can come in wet or dry forms.

ARGENTINA

■ EL MOLINO DEL MANZANO
Fondo de la Legua 375
1642 San Isidro
Buenos Aires
Tel: (54) 1 763 8682

Cotton Linters; Cotton Rag Half-Stuff; Sisal

Ready-to-Use Pulps (prepared pulps from plants, i.e. banana, cattail, etc.)

AUSTRALIA

■ PAPER CAPERS
De-Arne King
PO Box 281
Neutral Bay, NSW 2089
Tel: (02) 964 9741

Recycled Paper Pulp; Indigenous Fibres: fully prepared (must be picked up)

■ THE PAPER MERCHANT
316 Rokeby Road
Subiaco, WA 6008
Tel: (09) 381 6489
Fax: (09) 381 3193

Abaca; Cotton Linters; Eucalyptus (wood pulp); Sequoia (wood pulp)

BELGIUM

■ JEAN DECOSTER
Hannuitsesteenweg 114
B-3300 Tienen
Tel: (16) 81 12 86

Write for further information.

CANADA

■ KAKALI HANDMADE PAPERS, INC.
1249 Cartwright Street
Vancouver, B.C. V6H 3R7
Tel: (604) 682 5274

Abaca
1 Bleached Abaca: Creamy-white in colour.
2 Unbleached Abaca: Light beige in colour, with occasional strands of visible fibre.

Cotton Linters
1 Cotton Linters, 1st Cut: Recommended for sheet forming.
2 Cotton Linters, 2nd Cut: All-purpose pulp for sheet forming and casting.

Flax and Linen (bleached and unbleached)

Cotton, Silk or Linen Rag
1 Coloured and 2 White. 100% natural fibre content only, to be provided by the customer and cut (not torn) into 0.5in pieces. Up to 0.5lb of dry fibre hydrated in 5gal of water.

■ THE PAPERTRAIL
1546 Chatelain Avenue
Ottawa, Ont. K1Z 8B5
Tel: (613) 728 4669
Fax: (613) 728 7796
Orders: 1 800 363 9735

Raw Fibre (any), cut, cooked and beaten for 2 hours including:
Abaca
1 Unbleached Abaca (light beige in colour).
2 Bleached Abaca (cream colour).

Cotton Linters
1 Cotton Linters 1st Cut (recommended for sheet forming).
2 Cotton Linters 2nd Cut (paper casting). Can be combined with 1st Cut for extra strength.

Cotton Rag Pulp
Made from cotton rags which have been torn, cooked, beaten and made into pulp. Must be beaten in a Hollander. White or blue cotton rag available.

Bleached Flax Pulp
May be prepared in a kitchen blender but yields better results when beaten in a Hollander.

Also Raw Hemp and Hemp Half-Stuff; Raw Sisal and Sisal Half-Stuff; Prepared Pulps with a standard beating time of 1 hour for normal sheet forming. Additional beating on request. Coloured Pulps to customer specification; Sized Pulps with internal sizing and calcium carbonate added to customer specification.

(Illustration A rag collector)

■ LA PAPETERIE SAINT-ARMAND 3700 Saint Patrick Street Montreal, Quebec H4E 1A2 Tel/Fax: (514) 931 8338	Abaca; Cotton Rag; Cotton Linters
DENMARK	
■ AV-FORM Hammershusvej 14 c DK-7400 Herning Tel: (97) 22 22 33	Abaca; Cotton Linters
GERMANY	
■ EIFELTOR MÜHLE Auf dem Essig 3 D-53359 Rheinbach-Hilberath Tel: (22 26) 21 02	Cotton Linters (Bleached white) 1 Standard: slightly beaten. 2 Extra: beaten for 2 hours in a Hollander and can be easily rehydrated. 3 Flocks: requires beating in a Hollander. Cotton Linters (unbleached) 1 Flocks. Also Abaca; Spanish Hemp; Spanish Flax
■ DRUCKERN & LERNEN GmBH Bleicherstrasse 12 D-2900 Oldenburg Tel: (441) 163 34 Fax: (441) 140 88	Cotton Linters
■ PETER TEMMING AG Postfach 1220 D-25343 Glückstadt Tel: (41) 246 11 Fax: (41) 612 09	Cotton Linters

JAPAN

■ FUJI PAPERMILLS
C.P.O. Box 114
Tokushima 7790-34
Tel: (0883) 422 035
Fax: (0883) 426 085

Abaca; Wood Pulp (White, Sequoia)

(See also ORIENTAL PAPERMAKING)

SPAIN

■ CELESA
PO Box 76
E-43500 Tortosa
Tel: (77) 44 07 12
Fax : (77) 44 08 03

Linen; Abaca; Flax. Also other special fibres (Minimum quantity 210kg)

■ CELLULOSA DE LEVANTE
Apartado 346
E-08080 Barcelona
Tel: (3) 415 03 30
Fax: (3) 218 94 52

Abaca; Cotton Linters; Flax; Hemp; Sisal. (Larger quantities only, i.e., over 30kg)

■ ENCE
Av Burgos 8-B
E-28036 Madrid
Tel: (91) 337 8500
Fax (91) 337 8602

Eucalyptus pulp; Conifer Pulp (Minimum quantity 1000kg)

■ JULIO MONTFORT DELMAS
Doctor Balari 151-153
E-08203 Sabadell
Tel: (3)710 50 85
Fax: (3) 710 69 66

Linen; Abaca

■ MUSEU MOLI PAPERER DE CAPELLADES
E-08786 Capellades
(Barcelona)
Tel/Fax: (3) 801 28 50

Cotton Linters; Flax; Hemp; Sisal

SWEDEN

■ ZEN ART PAPER AB St. Jorgens Vag 20 S-422 49 Hisings Backa Tel: (31) 55 68 55 Fax: (31) 55 56 85	Cotton Linters; Cotton Rag; Raw Cotton
■ AB KLIPPANS FINPAPPERS-BRUK LESSEBO BRUK S-360 50 Lessebo Tel: (478) 106 00 Fax: (478) 107 58	Cotton Linters; Flax

SWITZERLAND

■ PAPIER ATELIER Teufenerstr .75 CH-9000 St Gallen	Jeans; Cotton; Straw
■ THERESE WEBER Atelierhaus, Fabrikmattenweg 1 CH-4144 Arlesheim Tel/Fax: (61) 701 91 07	Unbleached Abaca

UK

■ THE BRITISH PAPER COMPANY Frogmore Mill Hemel Hempstead, Herts HP3 9RY Tel: (01442) 231234 Fax: (01442) 252963	A new UK-based supplier of small quantities of papermaking requirements. Specialists in waste paper. Waste paper can be supplied as dry waste in bags or as half-stuff (with limited shelf life). There are many categories of waste paper; please contact mill for further details. Other Fibres Suppliers of small quantities of cotton linters and other special fibres and pulps. All enquiries welcomed.
■ FALKINER FINE PAPERS 76 Southampton Row London WC1B 4AR Tel: (0171) 831 1151 Fax: (0171) 430 1248	Cotton Linters

PAPER PLUS
24 Zetland Road
Chorlton, Manchester M21 2TH
Tel: (0161) 881 0672
Fax: (0161) 445 7273

Recycled Paper; Wood Cellulose (acid free); Plant Pulps

JOHN PURCELL PAPER
15 Rumsey Road
London SW9 OTR
Tel: (0171) 737 5199
Fax: (0171) 737 6765

Range of papermaker's half-stuff/pulps including Cotton Linter Pulp; Esparto Pulp; Brown Craft Pulp (long-fibred, unbleached wood pulp). Phone or fax for details.

USA

CARRIAGE HOUSE PAPER
79 Guernsey Street
Brooklyn, NY 11222
Tel/Fax : (718)599 PULP (7857)
Orders: 1 800 669 8781

Beaten pulps and special custom beatings (with no other beating work required).

Abaca
1 Unbleached Abaca: Light beige in colour, with occasional strands of visible fibre.
2 Bleached Abaca: Creamy white in colour, and the 'whitest' of the three types. It has the same qualities as the unbleached but has been bleached.
3 Premium Abaca: Creamy white in colour, but it has not been bleached. The Philippine government rigidly grades abaca by quality. This abaca is the highest grade and has a natural whiteness without bleaching.

Blue Denim Pulp
Made from 100% cotton rag beaten in a Hollander by Cave Paper and then dried.

Cotton Linters
1 Cotton Linters (1st Cut).
2 Cotton Linters (2nd Cut).

Unbleached Cotton Linters (therefore has some specks in it)

Cotton Rag Pulp

Bleached Flax Pulp

Linen Rag Fibre
100% linen scraps, in mixed colours. Must be beaten in a Hollander.

Sisal Pulp

■ DIEU DONNÉ PAPERMILL, INC.
433 Broome Street
New York, NY 10013
Tel: (212) 226 0573

Wet beaten pulp is pressed when wet and shipped damp for less costly shipping and easier storage. Pulp can be refrigerated for up to six weeks or dried out for storage. Damp pulp must be soaked in hot water and broken apart before reconstitution in a blender, hydro-pulper or paint mixer.

Bleached/Unbleached Abaca; Cotton Linters; 100% White Cotton Rag; 100% Coloured Cotton (when available); 100% Linen Rag Pulp (beaten for 4 or 8hr); 100% Flax

■ DOBBIN MILL
50-52 Dobbin Street
Brooklyn, NY 11222
Tel: (718) 388 9631

Custom Beaten Pulps

■ GOLD'S ARTWORKS, INC.
2100 North Pine Street
Lumberton, NC 28358
Tel: (919) 739 9605
Orders: 1 800 356 2306

Cotton Linters
1 Grade 8R (1st Cut): Long-fibre, cotton linter pulp. For sheet formation and works requiring strength of a longer fibre.
2 Grade 225 HL (2nd Cut): Short fibre. Good for casting, pouring, beginner's classes, etc. Breaks down quickly in a kitchen blender.
3 Grade 12R (Mill-run): Medium fibre length. Excellent all-purpose fibre.
4 Grade 100 U: Unbleached short fibre. Light tan with a 'natural' look. Limited availability.
5 Grade M-38: Staple cotton and cotton linter blend. 70% grade 8R, 30% staple fibre. Very strong. More like half-stuff without the problems of synthetic impurities.

Abaca
Light, off-white colour, excellent for combining with cotton linter pulp for additional strength or making thin sheets *Nagashizuki* style.

■ LEE S. McDONALD, INC.
PO Box 264
Charlestown, MA 02129
Tel: (617) 242 2505
Fax: (617) 242 8825

All beaten pulps are processed in a Hollander beater to customer specification. Overbeaten ready-to-use pulp for pulp painting may be ordered. Sizing or methyl cellulose can be added by request. Each 5gal pail contains one beater load of 2lb of dry weight of pulps with excess water drained off.

Cotton Linters
1 290 Cotton Linter (Mill-run): A mixture of 1st and 2nd cut linters.
2 282RS Cotton Linter (2nd Cut): One of the easiest pulps to use, it has been refined to form a good sheet after rehydrating. This pulp can also be used for sculptural papermaking.
3 282R Cotton Linter (2nd Cut): Also a 2nd cut linter, but longer and softer than the 282RS, as it has been less processed. A good choice for sculptural papermaking and with further beating will produce a crisp sheet.
4 270 Cotton Linter (2nd Cut): Another 2nd cut linter which is recommended for casting and sculptural use. Makes a bulkier and softer paper.

Abaca
1 Bleached Abaca: A light ivory-coloured abaca which can be used with other pulps.
2 Unbleached Abaca: A cream to tan-coloured abaca.

Spanish Flax
An off-white pulp with all the preliminary hours of beating and cooking already accomplished. Also known as Lincell B.

Hemp
This is a cream-coloured, long-fibred pulp which makes a very strong paper. It can be rehydrated in a blender, but needs to be beaten for best results. Also known as Hempcell B.

Sisal
1 Sisal: A cream-coloured pulp which rehydrates well.
2 Unbleached Sisal: Slightly darker in colour than ordinary sisal. Some sheets have clearly-visible fibres embedded in them.

Pulp Sample Kit: Contains one sheet each of the dry pulps.

■ MAGNOLIA EDITIONS
2527 Magnolia Street
Oakland, CA 94607
Tel: (510) 839 5268
Fax: (510) 893 8334

A 5 gallon minimum order is required for beaten pulps which are shipped in buckets with locking lids. Pulp can be strained to reduce moisture content (for an additional charge).

Abaca
1 Abaca: Premium White
2 Bleached Abaca
3 Unbleached Abaca

Cotton Linters
1 Cotton Linters (1st Cut)
2 Cotton Linters (2nd Cut).

Cooked *Kozo*
This fibre has been soaked overnight and cooked for 2 hours. The bark has not been scraped or picked clean. The amount is based on the dry weight of the fibre.

Also Cotton Rag; Spanish Flax; Hemp; Casting Pulp

■ GREG MARKIM INC.
PO Box 13245
Milwaukee, WI 53213
Tel: (414) 453 1480
Fax: (414) 453 1495
Orders: 1 800 453 1485

Pulps are described as 'shredded' and include Cotton Linters for papermaking or casting; Blue Denim and Cotton Rag Pulp; Green Cotton and Wool Pulp; Grey Cotton and Wool Pulp; Scrubbed Corn Husk and Cotton Pulp; White 100% Cotton Rag Pulp; Plum Cotton Rag Pulp with Flecks; Mixed Petals for inclusions.

- **PYRAMID ATLANTIC**
6001 6th Avenue, Suite 103
Riverdale, Maryland 20737
Tel: (301) 459 7154/(301) 577 3424

Custom Beaten Pulps in 5gal containers include: Abaca; Cotton Linters; Cotton Rag

- **THE PULPERS**
1101 North High Cross Road
Urbana, IL 61801
Tel: (217) 328 0118

Sheet Forming Pulp
A special blend of cotton, linen and hemp fibres, chosen for their strength and ease of handling. Available in 2 sizes: 2 or 5gal containers.

Casting Pulp
Available 2 or 5gal containers.

Cotton Linter and 100% Cotton Rag Pulps
Both made from neutral pH fibres free from any chemicals.
1 100% Cotton Linter Pulp: Short-fibred, will produce a soft, absorbent paper. It is a fast draining and a less expensive 'beginners' pulp for children to learn paper casting and sheet forming. In 2 or 5gal containers.
2 100% Cotton Rag Pulp: Long-fibred and will produce a strong, white paper. It is slow draining and a favourite with experienced papermakers. In 5gal containers.

- **SEA PEN PRESS AND PAPER-MILL**
2228 N.E. 46th Street
Seattle, WA 98105
Tel: (206) 522 3879

Custom Beaten Pulp (5gal containers); Cheney Half-Stuff; Abaca; Linen

- **TWINROCKER HANDMADE PAPER**
PO Box 413
Brookston, IN 47923
Tel: (317) 563 3119
Fax: (317) 563 8946

Custom Beaten Pulps
This is one of Twinrocker's specialities. The range changes and widens so please contact the mill for current information. A wide range of fibres (see before) and beaten ready-to use pulps. All pulps are beaten to customer specification, according to intended use, i.e., casting, sheet forming, etc. Pulp is shipped in 5gal plastic pails, each of which holds one batch (3lb dry weight) with two-thirds of the water removed. Special price reduction for groups of 5 batches of a single pulp. Orders and enquiries welcome by telephone.

Cotton Rag TR 89
This is a bleached muslin that is warm white in colour.

Black Cotton Rag
Black cotton from the clothing industry.

Esparto
Although esparto fibres are relatively short and not very strong, the pulp shrinks very little in drying and is, therefore, good for casting work.

Linen Rag Yardage
Khaki-coloured, fabric scraps woven from combed line flax. This makes a paper that is slightly translucent, rattly, hard and strong, and tends to shrink a great deal in drying.

Also Dried Flowers and Plants

■ WOMEN'S STUDIO WORKSHOP
PO Box 489
Rosendale, NY 12472
Tel: (914) 658 9133

Abaca

Cotton Linters
(Both available in 5gal containers.)

Recycled pulps

Recycled paper pulp offers an alternative source of pulp which is to many worth considering. Paper has a long history as a recycled material and today many artists believe in using waste papers as part of a wider environmental policy. Industrial and commercial papers provide a readily available supply, although they should not be used indiscriminately. If high standards are required, the best papers for recycling are quality handmade or mould-made papers. (See PAPER CAPERS, THE BRITISH PAPER COMPANY and PAPER PLUS.)

Recycling of paper must be part of any resolution to the current garbage crisis . . .

Claudia G. Thompson *Recycled Papers - The Essential Guide* 1992

Three hundred papermills or more
make valleys echo with their tread
like giant armies marking time. . .

. . . to lift a row of ponderous stakes
then lets them fall like angry fists
in iron plated mortar bowls.

From a 17th century poem, *Papyrus* by Father Imberdis

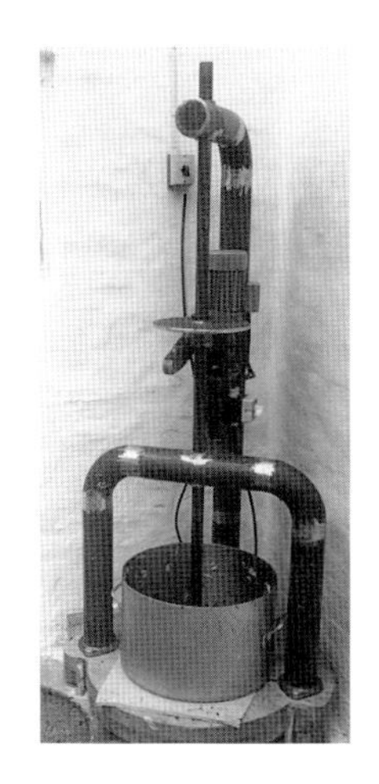

Stampers, Hollander beaters, Pulp processors

Above   Island Stamper, Anne Vilsbøll
Right   7lb Hollander Beater, David Reina Designs
Bottom   Wooden stampers

Stampers

Wooden stampers at the Richard de Bas Mill, France

DENMARK

■ ANNE VILSBØLL
Fredensgade 4
Strynø
DK-5900 Rudkøbing
Tel: (62) 51 50 02
Fax: (62) 51 50 12

Island Stamper
Designed by Flemming Engedal Andersen and built by Frede Lundsgaard, 1987-88.

(See over for illustration)

USA

■ LEE S. McDONALD, INC.
P.O. Box 264
Charlestown, MA 02129
Tel: (617) 242 2505
Fax: (617) 242 8825

Pneumatic Stamper
Plan, drawings and instructions for building a pneumatic stamper, driven by compressed air.

. . . Joel Fisher once starved himself for five days, then masticated paper, including a bank note, and formed art as a result . . .

Hollander beaters

AUSTRALIA

■ THE PAPER MERCHANT
316 Rokeby Road
Subiaco, WA 6008
Tel: (09) 381-6489
Fax: (09) 381-3193

Hollander Beaters
Custom made to specifications.

CANADA

■ FLAX PAPER WORKS
Helmut Becker
R.R. 4,
Komoka, Ontario
NOL 1RO
Tel: (519) 666 0624
Fax: (519) 679 2443

Mississauga Beaters
Corrosion-resistant, all-stainless-steel assembly of beater roll, shaft, flybars, bedplate bars, backfall, cover and tub. Variable speed motor control with safety switch and automatic overload shut off. Precision settings assure quality control of hydration, cavitation, maceration and fibrillation of all plant fibre stuff from strongest line flax to raw cotton. Manufactured in Mississauga, Canada, by College Tool and Die Ltd., a firm specialising in prototypes and high-technology parts for the aircraft, energy and space industries. Three sizes: 1 1lb Model 2 3lb Model (new design) 3 5lb Model. Contact Helmut Becker.

GERMANY

■ EIFELTOR MÜHLE
Claudia Stroh Gerard
Auf dem Essig 3
D-53359 Rheinbach-Hilberath
Tel: (22 26) 21 02

Peter Beaters (see PETER GENTENAAR; THE NETHERLANDS); Also a Pulper. Contact the mill for further information.

INDIA

■ BRAMCO ENGINEERS
E-55-56 Industrial Area
Yamuna Nagar
135001 Haryana
Tel: (1732) 22231
Fax: (1732) 26610

Hollander beaters in stainless steel, mild steel or combinations from 0.5kg (1lb) to 1.3kg and 5kg up to full production scale (5 ton). Small beaters can be made to a very sturdy specification, e.g. from 7mm stainless steel plate complete with tub, motor, drum washer, stand, etc. Production beaters usually have a concrete tub made by the customer to drawings supplied.

Bramco also make hydropulpers, centrifugal cleaners, rag cutters, presses, glazing rolls, cylinder-mould machines, etc. As with most Indian suppliers, prices are low but allow plenty of time for correspondence and delivery. Be very precise about materials and specifications.

■ UNIVERSAL ENGINEERING CORPORATION
Sabri Mills
Ambala Road
Saharanpur-247001 (UP)
Tel: Off: 8427 Res: 4221 6427
Fax: (132) 5318

Universal Laboratory Beaters (Valley Type)
Tub lined with stainless steel. Beater roll mounted in ball bearings in the tub. Beater roll and bedplate made of special non-rusting alloy. Controlled bedplate pressure. Capacity tub volume 25ltr / 680g bone dry pulp at 2 to 2.5% concentration. Designed in accordance with TAPPI & SCAN standards.

Universal Hollander Beater 1 Capacity: 2000g; 2 Capacity: 3000g; 3 Capacity: 5000g.

THE NETHERLANDS

■ PETER GENTENAAR
Churchilllaan 1009
NL-2286 AD Rijswijk
Tel: (1742) 96961

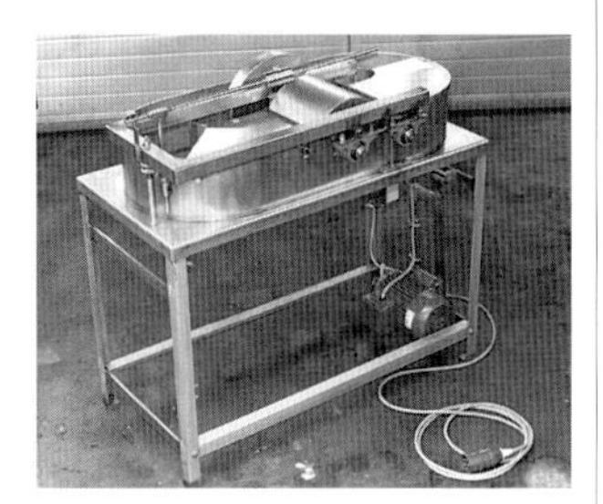

The Peter Beater

The Peter Beater
A recently designed beater which processes fibres such as flax, sisal, jute, hemp and ramie as well as partially prepared pulps. All stainless steel with hinged roll and curved bronze bedplate for effective beating. Adjustable distance between roll and bedplate: sliding weight for adjustable pressure on the roll. Capacity: 3ltr; Dry fibre capacity: 500g.

. . . history is not a science, it is an art, and man succeeds in it only by imagination. . . Anatole France

PHILIPPINES

■ BERNARD HANDMADE PAPER PRODUCTS
Concepcion, San Pablo City
Philippines 4000

Grass Cutter/Paper Tearing Machines
This machine has been adopted for cutting plant materials or tearing used papers into digestible size in preparation for chemical treatment. Frame and housing is made of iron and steel. Capacity: 50kg of material per hour.

Pulp Cookers
For boiling plant materials during the chemical treatment phase. Made of stainless steel sheet with a capacity of 40 litres of pulp/chemical solution.

Beater Machines
For pulping and homogenising materials before loading onto moulds. The drum is made of stainless steel with tool-steel beater blades. The drum is tiltable for ease of handling, with a capacity of 40 litres of pulp/chemical solution.

Stoves
For cooking pulp with a steady supply of water and ready air for the dryer. Made of concrete hollow block and adobe with cement and lime. Fitted with water heater pipes.

Sieve Boxes
For the washing of chemically-treated materials after boiling. Made of oiled wood and brass screen/mesh. 20 x 5in high.

Screen Boxes
Placed at the top of the sieve box, for catching the pulp while separating and washing the material after chemical treatment. Oiled wood and brass mesh.

SPAIN

■ TALLERES SOTERAS
Amador Romani, s/n 08786
E-Capellades Barcelona
Tel: (3) 801 02 05
Fax: (3) 801 30 06

Designer and manufacturer of equipment especially for the paper industry. 5 sizes of beater available beating anything from 0.2 to 240kg of pulp. Write for more details and a price list.

UK

■ MESSMER INSTRUMENTS LTD.
Unit F1 Imperial Business Estate
West Mill
Gravesend, Kent DA11 0DL
Tel: (01474) 566488
Fax: (01474) 560310

Precision beaters which will subject repeated samples to a uniform, reproducible and stable treatment over substantial periods of use. Beating pressure is exactly reproducible by means of the controlled bedknife.

5lb Valley Laboratory Beater
The interior of the tub, roll billet and bedknife box is of cast-iron construction. The flybars and bedknife are of 410 stainless steel, in accordance with TAPPI specifications. Complete washer cylinder equipment is available for 1.5lb beaters to provide a means of washing the stock after bleaching (or whenever washing is required).

USA

■ CARRIAGE HOUSE PAPER
79 Guernsey Street
Brooklyn, NY 11222
Tel/Fax: (718) 599 PULP (7857)
Orders: 1 800 699 8781

David Reina Beaters (see over)

■ DAVID REINA DESIGNS INC.
79 Guernsey Street
Brooklyn, New York 11222
Tel: (718) 599 1237
Fax: (718) 599 7857

(see p. 33 for illustration)

David Reina Designs specialise in making high-quality, studio-sized beaters. They supply three different traditional Hollander beaters for the needs of private studios, schools and small production mills, repair antique beaters and also make custom equipment for paper-making.

David Reina Beaters
Two options are available for both models: Castors and Digital Height Indicator (which shows the precise space adjustment between the roll and bedplate).
1 Stainless Steel Model: The standard model beater has a dry pulp capacity of 2lb. It features a stainless steel tub, bedplate, shaft, cover and roll blades. The blades are set into a bronze spool.
2 Stainless Steel/Aluminium Model: This beater will also beat 2lb of dry pulp. It features an aluminium roll machined from a solid block of high-strength aluminium, with a stainless steel tub and chassis.
3 7lb model: This beater has been designed and built specifically for the laboratory of a major paper company. Stainless steel is used for the tub, bedplate, shaft, cover and bars. The bars are set into a bronze roll.

■ LEE S. McDONALD, INC.
P.O. Box 264
Charlestown, MA 02129
Tel: (617) 242 2505
Fax: (617) 242 8825

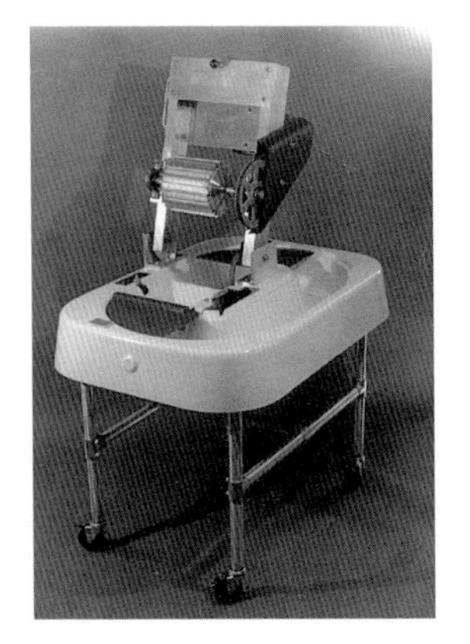

New Hydra Hollander Beater

New Hydra Hollander Beaters
Especially designed with safety, the ability to make good pulp and economy in mind. These beaters can be used to beat partially-processed pulps, fibres such as flax and certain indigenous fibres. They hold 2lb of pulp or raw fibre. The cast 304 stainless steel beating roll is mounted on a stainless steel shaft with a matching, grooved, stainless steel bedplate. The beater tub is made of reinforced polyester resin. An optional amp meter is offered which allows the user to keep an accurate record of beater cycles by measuring the amp set when beating. A clear plexiglass lid, which covers the full opening of the beater and helps lower the noise level, can also be supplied.

New Hydra Hollander Beater for International Papermakers
Dimensions are 35 x 49 x 48in high.

4lb Hollander Beater
Design based on the New Hydra Hollander Beater.

Pulp processing equipment

CANADA

- CRANBERRY MILLS
R.R. No. 1
Seeleys Bay,
Ontario KOH 2NO
Tel: (613) 387 1021

'Do-it-yourself' Hydropulpers
Drawings and equipment list to construct a 'do-it-yourself' hydropulper with box dimensions 36 x 36in square x 45in high, with a 2 x 2in angle iron frame and agitator. Contact Edward H. Snider.

Angle Iron Frames
Drawings to construct an angle iron frame on which to mount a Hollander beater to permit circulation of pulp from the pulper to the beater and back again. This enables a 1lb beater to be converted to a 10lb to 20lb beater.

- KAKALI HANDMADE PAPERS, INC.
1249 Cartwright Street
Vancouver, B.C. V6H 3R7
Tel: (604) 682 5274

Pulp Whizzers
Used for hydrating pulp with the aid of regular drills.

USA

- CARRIAGE HOUSE
9 Guernsey Street
Brooklyn, NY 11222
Tel/Fax: (718) 599 PULP (7857)
Orders: 1 800 699 8781

Mixer Blades
Great for preparing pulp or for mixing pigments; all stainless-steel blades fit into US variable-speed, electric hand drill. Various sizes.

- LEN KEL MANUFACTURING
PO Box 2996
San Rafael, CA 94920
Tel: (510) 232 0949

The Garrett Hydropulpers and Mixers (Studio model)
The first real alternative to the kitchen blender versus the Hollander Beater. Safe, convenient, compact, quiet and efficient unit which rehydrates partially processed pulps in a few minutes. Includes a pulp separator to strain out the pulp, which can also be pumped out directly into a vat, etc. Holds 1lb of fibre per 20gal of water. The Hydropulper can also be used for evenly dispersing pigments, uniformly blending dyed pulp or additives, etc.

The Processor (Studio model)
This system can process bast fibre (*kozo*, etc.), native fibre and sized paper. Water recycling pulp separator included with every unit. Holds 1lb fibre/20gal of water.

The Rehydrator (Portable and Studio model)

■ GOLD'S ARTWORKS, INC.
2100 North Pine Street
Lumberton, NC 28358
Tel: (919) 739 9605
Orders: 1 800 356 2306

The Processor (Studio model) (see LEN KEL MANUFACTURING)

The Garrett Hydropulper (Studio model) (see LEN KEL MANUFACTURING)

■ LEE S. McDONALD, INC.
PO Box 264
Charlestown, MA 02129
Tel: (617) 242 2505
Fax: (617) 242 8825

The Whiz Mixer
An economical rehydrator made for a wide variety of commercially available pulps and for reconstituting custom prepared pulps. Can also be used to blend colours, sizing, retention agent, and other additives into the pulp. Not designed to process raw fibres. The 4in cast stainless steel propellor attached to a 36in long, 0.5in diameter stainless steel shaft can properly hydrate 1 to 3lb of pulp at a time in a 20 to 30gal plastic container. Swivel mount for vertical or horizontal mounting. Spare parts available.

■ MAGNOLIA EDITIONS
2527 Magnolia Street
Oakland, CA 94607
Tel: (510) 839 5268
Fax: (510) 893 8334

Garrett Hydropulper and Mixer (Studio model) (see LEN KEL MANUFACTURING)

The Processor (Studio models) (see LEN KEL MANUFACTURING)

The Rehydrator (Studio and portable models) (see LEN KEL MANUFACTURING)

KITCHEN BLENDERS

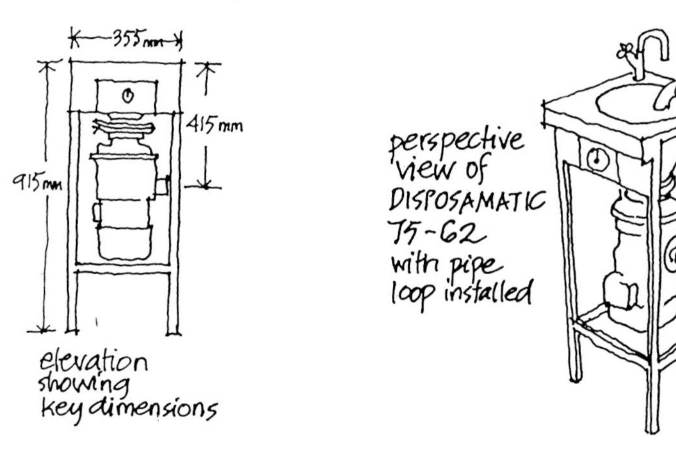

■ PLANT PAPERS
Romilly, Brilley
Herefordshire HR3 6HE
UK
Tel (01497) 831 546
Fax (01497) 831 327

"Many people wanting to get into papermaking start by making pulp in their kitchen blenders, which have limitations. Their action is a cutting rather than a bruising one and the flasks are small" writes Maureen Richardson of Plant Papers in England. The leap in cost from cheap domestic equipment to the smallest Hollander beater on the market is prohibitive for the small producer. Maureen has converted a sink waste-disposal unit, which although it does not do everything a Hollander will, is a sturdy, reliable and cost-effective machine. She has published a small booklet, *An Alternative to the Beater?*, describing her new adaptation of a Disposamatic 75-62 free-standing model, produced in the UK for catering and laboratory use. With her permission we reproduce drawings of the converted machine. Contact Plant Papers for more information.

Pigments were created from exquisite and bizarre ingredients: blue from lapis lazuli and a colour called Indian Yellow made from the urine of a cow which had been fed on mangoes and made to stand in the sun. . .

Nigel Macfarlane *A Paper Journey* 1993

Chemistry in papermaking
Additives and adhesives
Pigments

Chemistry in papermaking

Many characteristics of a finished sheet of paper are determined by the use of chemical additives during its making. For example, paper can be dyed, filled and coated; it can be made water-resistant; it can have exceptional tensile strength or virtually none at all.

Of the many chemical additives employed in papermaking, perhaps the two most often used are sizes and pigments. Sizes are employed to control wetting of the sheet and the ability of paper to accept ink or other water-based fluids without undue bleeding or feathering. Traditional sizes used in Western papermaking were made of a gelatine solution, into which the already dried sheets were immersed (external sizing). Sizes added either at the beating stage or to the vat (internal sizing) were developed in the late 18th century, principally alum and rosin. Alum is also used as a mordant or colour fixative, but its high acidity causes irreversible damage and it is not a particularly beneficial additive. The synthetic sizes now used by papermakers are less acidic and help produce paper with a neutral pH range. Today most additives are put into the pulp either at the end of the mixing and/or beating cycle. Most colorants, for example, are now mixed in with the pulp before the paper is made and, in the case of pigments, require an additive known as a 'retention aid' to help them adhere to the fibre.

The *tororo-aoi* plant

Other additives used in papermaking include fillers (which reduce shrinkage during drying and impart a smoother surface) and whiteners (which increase brightness and opacity). Some, like calcium carbonate, also provide an alkaline reserve in the paper which promotes longevity. However, because they end up in the space between the fibres, these additives may hinder bonding between the fibres and lessen the strength of the finished paper. Conversely some additives, for example, methyl cellulose powder, can be added to the pulp to promote fibre-to-fibre bonding. This increases the strength of a short-fibred paper, as for example, in paper castings.

Papers made during October and November will mature quickly and size easily. This is due to the foggy mornings in the late Autumn which soften the waterleaf paper in the loft after it is dried, if the louvre boards are fully open.

J Barcham Green *Papermaking by Hand in 1953* (Out of prir

Finally, some natural additives have a remarkable ability to control sheet formation. One such ingredient in Japanese papermaking is *neri*, a mucilage which is extracted from the root of the *tororo-aoi* plant (a member of the hibiscus family) and a number of other plants. It is also known as a formation aid (or deflocculant) because it helps to separate the long bast fibres in the vat and slows the drainage of the water through the mould or *su*. Synthetic formation aids are frequently used with European-style papermaking to assist the distribution of fibres and achieve various decorative effects.

Additives and adhesives

A list of the more widely used additives and adhesives includes

CALCIUM CARBONATE

A pure form of limestone that has been ground into a powder. It is used as buffering agent and is added to the pulp to protect the paper from any future acidity that may be present in the atmosphere. When used in larger amounts it acts as a filler to improve opacity and whiteness and can be used to retard shrinkage in paper castings.

FORMATION AID

Used in Japanese papermaking as a deflocculant for long-fibred pulps; it keeps the fibres from entangling and allows the mould to be dipped repeatedly in the vat to form thin, translucent sheets. Can also be used for pulp painting and decorative papermaking techniques. Several products are available as a synthetic substitute for *neri*, in either powder or liquid form.

1 PMP is cationic (positively charged) and is generally used in *Nagashizuki* (Japanese-style) papermaking. It is essential when using Japanese fibres and can be used with abaca to form a thinner sheet.

2 PNS is anionic (negatively charged) and can used as a formation aid when no other additives are present. However, when PNS is used with coloured pulps, the fibres will clump or coagulate so that individual coloured fibres remain distinct when mixed together in the same vat.

INTERNAL SIZING

Usually available as an alkene-ketene-dimer emulsion that can be used in neutral pH conditions and meets archival standards. Makes paper less absorbent, so that ink or paint does not bleed or feather.

KAOLIN (China clay)

A very fine white powder (hydrous silicate of aluminium) that can be added to the pulp as a filler to reduce shrinkage and is often used as an additive for sculptural techniques. It also leaves a smoother surface to the paper.

METHYL CELLULOSE (*Sodium carboxy methyl cellulose*)

A versatile adhesive in powder form which can be used for strengthening short-fibred pulps; especially useful when working with paper castings. It adds some sizing (water resistance) characteristics and also makes the surface of finished art pieces more durable.

RETENTION AID

A cationic agent, available in powder or liquid form, used for the purpose of retaining or holding pigments to the pulp fibres. It ensures a colour's full intensity and virtually eliminates the need to rinse the pulp.

SODA ASH (*Sodium carbonate*)

An alkaline substance used in the cooking of fibres to dissolve the non-cellulose parts which are detrimental to the papermaking process.

TITANIUM DIOXIDE

A very fine, white powdered pigment mainly used to add brightness and opacity to paper. Like calcium carbonate and kaolin, it also acts as a filler, giving a smoother surface to the paper and resulting in less 'pick' (see also PIGMENTS).

Pigments

Coloured paper can be made using many different methods, but most hand papermakers use dyes or pigments to colour their pulp. Dyes are water-soluble colouring agents that have natural affinity to cellulose and penetrate the structure of the fibre in order to become attached. They are usually added to the raw material before beating. Few dyes are entirely colourfast, lightfast or bleedproof when used alone and, in general, pigments provide a more permanent, even colouring. Pigments are insoluble, finely ground particles and must be attached to the fibre using a binder or 'retention agent', a small amount of which is added to the pulp before adding the pigment. Most hand papermakers use a synthetic binder which gives the fibre a positive charge to attract the negatively charged pigment. A range of aqueous-dispersed pigments (dispersed in concentrated form in water), specifically formulated for the colouring of paper pulp, together with the necessary retention agent, are widely available and come with full instructions for use. They are usually added to already beaten pulp before it is added to the vat. Pigments may be mixed to create greens, oranges, etc. before adding to the pulp, or previously coloured pulps can be mixed together just like paint. A range of iridescent and reflective pigments are also available. They can be added to coloured or uncoloured pulp and are used in conjunction with a retention aid. Metallic colours produce metallic shades and effects without the use of actual metallic flakes. They will not tarnish or degrade the finished paper.

NOTE ON HEALTH AND SAFETY

Many chemical additives can be quite toxic and are poisonous if ingested or inhaled. Some can cause skin irritations. Precautions are, therefore, urged in dealing with any chemical compound. Good ventilation and the use of a rubber mask are recommended when using powdered pigments; splash goggles and rubber gloves should also be used when mixing stock solutions. The use of a skin guard (barrier cream) may also be appropriate.

AUSTRALIA

■ THE PAPER MERCHANT
316 Rokeby Road
Subiaco, WA 6008
Tel: (09) 381 6489
Fax: (09) 381 3193

Formation Aid; Internal Sizing; *Konnyaku* (see ORIENTAL PAPERMAKING); Lick-n-Stick Envelope Glue: Low-moisture Paper Cement; Methyl Cellulose

Aqueous-dispersed Pigments; Hot and Cold Water Dyes; Acrylic-based Inks

CANADA

■ KAKALI HANDMADE PAPERS, Inc.
1249 Cartwright Street
Vancouver, B.C.
Tel: (604) 682 5274

Coagulant (see Formation Aid above); Formation Aid; Magnesium Carbonate; Methyl Cellulose; Retention Aid; Sizing; Soda Ash

Aqueous-dispersed Pigments; Regular Pigment Starter Kits

■ LA PAPETERIE SAINT-ARMAND
3700 Saint Patrick Street
Montreal, Quebec H4E 1A2
Tel: (514) 931 8338

Luster Pigments; Luster Pigments Starter Kits; Sandoz Direct Dyes

■ THE PAPERTRAIL
1546 Chatelain Avenue
Ottawa, Ont K1Z 8B5
Tel: (613) 728 4669
Fax: (613) 728 7796
Orders: 1 800 363 9735

Calcium Carbonate; Coagulant; Envelope Glue

Aqueous-dispersed Pigments; Sample Set of Pigments (12 colours); Powdered Pigments (imported from Europe, impressive range of colours and mostly coloured earths with some synthetics) may be mixed in small batches in a kitchen blender with retention aid or methyl cellulose to bind them to the fibre; Sample Set of Powdered Pigments; Luster Pigments; Sample Set of Luster Pigments; Italian Marble Dust (brown red Pugna and coral pink); Glitter (small polished and coated pieces of aluminium foil which are just added to your vat)

GERMANY

■ EIFELTOR MÜHLE
Auf dem Essig 3
D-53359 Rheinbach-Hilberath
Tel: (22 26) 2102
Fax: (22 26) 2102

Calcium Carbonate; Formation Aid; Internal Sizing; Kaolin; Potash; Retention Agent; Starch (a special cold-water starch derivative); Titanium Dioxide

Pigments

PHILIPPINES

■ JARNET TRADE INTERNATIONAL
Rm. 304 Donya Narcisa Bldg.
8751 Paseo de Roxas
Makati
Metro Manila
Tel: (2) 817 0706/(2) 817 1643
Fax: (2) 817 0709

Retention Agent

Dyestuffs (Distributor for Pergasol Liquid Direct Dyes with a range of colours)

SWITZERLAND

■ SANDOZ PRODUCTS (SCHWEIZ) AG CH-Lichtstrase 35 4002 Basle	Write for details.
■ THERESE WEBER Atelierhaus Fabrikmattenweg 1 CH-4144 Arlesheim Tel/Fax (061) 70191 07	Pigments

UK

■ THE BRITISH PAPER COMPANY Frogmore Mill Hemel Hempstead, Herts HP3 9RY Tel: (01442) 231234 Fax: (01442) 252963	Sizing Emulsion; Acids and Sodas; Starch; Retention Aid; Chalk; Clay Pigments
■ HERCULES LTD. 31 London Road Reigate RH2 9YA Tel: (01737) 242434 Fax: (01737) 224288	Internal Sizing Aquapel 360X ketene-dimer emulsion. A highly efficient sizing agent for use against a wide variety of penetrants. It is not dependent on alum and reacts directly with cellulose to provide sizing. Large quantities only. Retention Aid For maximum efficiency, Aquapel 360X should be supplemented with a small amount of a cationic retention aid (Kymene 557H, Kymene 709 or DEFLOC 510). Large quantities only.
■ SANDOZ CHEMICALS (UK) LTD. Calverley Lane Horseforth, Leeds LS18 4RP Tel: (01532) 584646 Fax: (01532) 390063	Dyes and chemicals aimed directly at the machine production of paper products but also applicable to handmade papers, including cationic and anionic dyes, retention agents and detergents for washing felts.

There is something both moving and significant in the reflection that blue paper once came from blue rags, probably sailor's clothes.

John Krill *English Artists Papers* 1987

USA

■ CARRIAGE HOUSE PAPER
79 Guernsey Street
Brooklyn, NY 11222
Tel/Fax: (718) 599 PULP(7857)
Orders: 1 800 699 8781

Calcium Carbonate; Coagulant; Formation Aid; Internal Sizing; Kaolin; Methyl Cellulose; PVA (Polyvinyl Acetate); Retention Soda Ash; Titanium Dioxide

Aqueous-dispersed Pigments; Glitter; Sample Set 12 Pigments; Luster Pigments; Sample Set of Luster Pigments

■ DIEU DONNÉ PAPERMILL, INC.
433 Broome Street
New York, NY 10013
Tel: (212) 226 0573

Internal Sizing; PVA; Retention Aid

Pigments

■ GOLD'S ARTWORKS, INC.
2100 North Pine Street
Lumberton, NC 28358
Tel: (919) 739 9605
Orders: 1 800 356 2306

Aqueous Acrylic Medium; Calcium Carbonate; Hyplar Hygel; Methyl Cellulose; Sandoz Retention Agent; Stalok Sizing (a pre-cooked potato-starch size in powder form which produces stronger bonding, increases strength and improves sheet surface).

Perma Colour Pigments: lightfast and chemically pure (free of fillers, extenders and dilutants) for excellent performance; Pearlised Metallic/Luster Dry Pigments.

■ LEE S. McDONALD, INC.
PO Box 264
Charlestown, MA 02129
Tel: (617) 242 2505
Fax: (617) 242 8825

CAB (*Cellulose acetate butyrate*)
An archival stiffener in powder form which also penetrates and protects paper. It is water repellent when properly applied to dried paper and will not leave a surface sheen, although it may darken the paper slightly. CAB must be mixed with acetone, and therefore should be used with caution. Test before using as acetone will leach out certain colouring agents.

Re-Lick Glue
A remoistenable, transparent adhesive for envelope flaps with a shelf life of 6 months.

Polyadam Polymer
Adds strength and stiffness to castings or assembled works of paper, fibre or cloth. Used mostly on dry paper as a surface coating, it can also be mixed into wet pulp. Pulp with Polyadam Polymer added can be worked into three-dimensional moulds or over armatures. It is a two-part vinyl co-polymer which must be mixed together in equal parts. Dries clear, flexible and waterproof but will leave a sheen when applied with a brush.

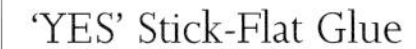

'YES' Stick-Flat Glue
An all-purpose vegetable paste which contains corn dextrine, corn syrup, water and a small amount of preservative with a pH of 6.0. Highly concentrated and needs thinning with water.

Also Calcium Carbonate; Formation Aid; Internal Sizing; Kaolin; CMC Methyl Cellulose (*sodium carboxy methyl cellulose*); Methyl Cellulose (Dow Methyl Cellulose A 4C); PVA Adhesive (a water soluble, polyvinyl acetate, dries clear with a slight sheen; archival and non-yellowing, withstands freezing temperatures); Retention Aid; Soda Ash; Titanium Dioxide C

Aqueous-dispersed Pigments; Sample Kits; Pearlescent Colours; Iridescents; Pearlescent Sample Kits; Glitter

■ MAGNOLIA EDITIONS
2527 Magnolia Street
Oakland, CA 94607
Tel: (510) 839 5268
Fax: (510) 839 8334

Calcium Carbonate; Coagulant PNS; Envelope Glue; Formation Aid; Internal Sizing (Hercon 70); Kaolin; Label and Foil Adhesive (holds its tack indefinitely; excellent alternative to gelatin size for gold leafing); Methyl Cellulose (hydroxypropyl methyl cellulose); Pritt Glue Stick (a neutral pH water soluble glue stick); PVA (this white glue is commonly used by bookbinders because of its flexibility and durability).

Aqueous-dispersed Pigments; Glitter; Pearlescent Pigments

■ PRO CHEMICAL AND DYE, INC.
PO Box 14
Somerset, MA 02726
Tel: (508) 676 3838
Fax: (508) 676 3980

PRO Retayne (Cationic dye fixing agent used on cotton fabrics to improve wet fastness of direct dyes and used with pigment colour concentrates in colouring paper pulp); Soda Ash

PRO Colour Concentrates; PRO Diazol Dyes; PRO Fiber Reactive Dyes

■ TALAS
568 Broadway
New York NY 10012
Tel: (212) 219 0770
Fax: (212) 219 0735

A supply house specialising in archival and conservation materials. Write for more details.

■ TWINROCKER HANDMADE PAPER
PO Box 413
Brookston, IN 47923
Tel: (317) 563 3119
Fax: (317) 563 8846

Calcium Carbonate; Formation Aid; Internal Sizing; Kaolin; CMC Methyl cellulose; Permanent Glue; Remoistenable Glue; Retention Agent; Soda Ash; Titanium Dioxide

Aqueous-dispersed Pigments; Crystal Metallic Fibre (very fine shredded metallic and iridescent mylar film in 14 colours); PRO Colour Concentrates; PRO Diazol Dyes; PRO Fiber Reactive Dyes; Pigments Sets (full = 13 colours); Pigment Sampler Sets (one of each pigment excluding white); Pearlescent or Luster Pigments; Pearlescent Sampler Set

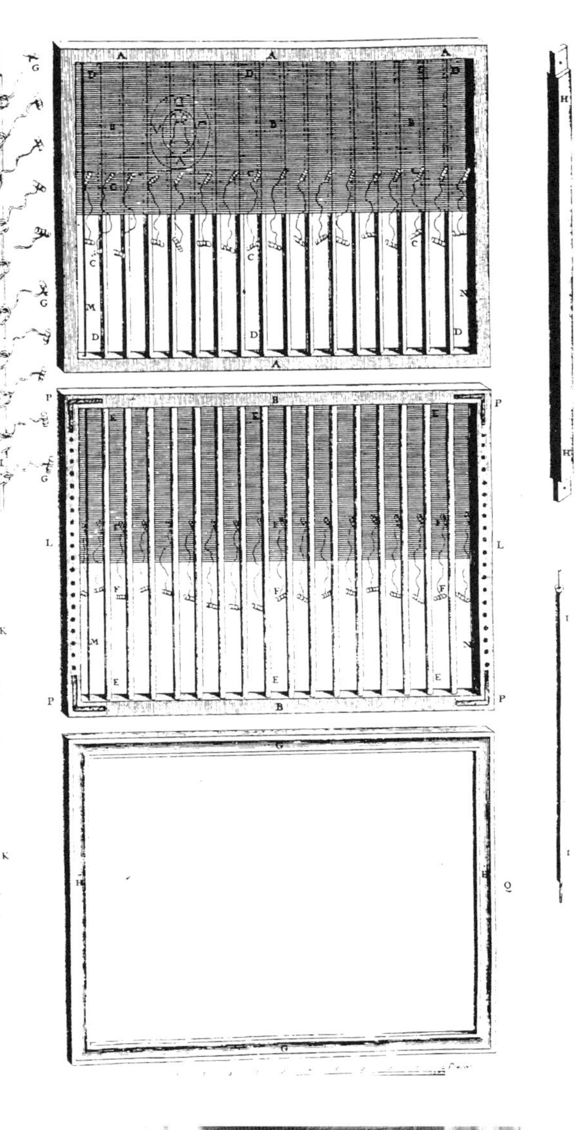

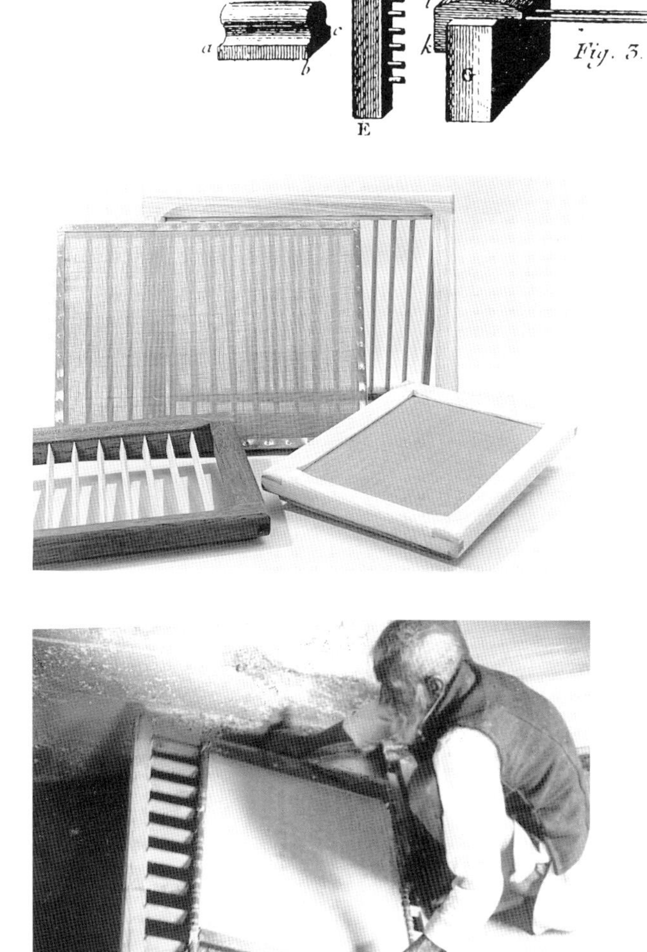

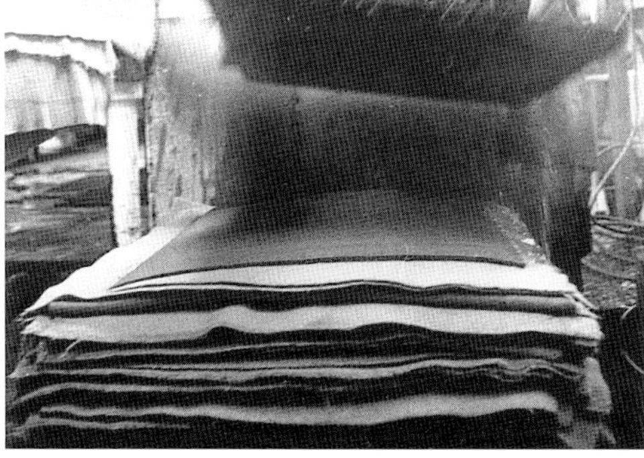

Moulds and deckles, felts
Mouldmaking components
Watermarks
Vats

Far top right Mould making components
Top right Hand-crafted moulds and deckles from Eifeltor Mühle
Right Lifting a sheet from a vat, Rajasthan
Top left Mould making
Bottom left Wet sheets interleaved with felts

Moulds and deckles, felts

Handcrafted moulds and deckles from Steam Film Pty Ltd.

The making of a Japanese *su*

A mould and deckle is the basic tool used by every hand papermaker to form a sheet of paper. Although its construction varies slightly according to particular sheet forming methods, its purpose remains the same: to allow the water to drain away from the thin deposit of fibres which settle on the surface of the mould during sheet forming. The mould is made of a wooden frame to which a fine mesh surface is secured. A separate, uncovered frame, the deckle, fits exactly over the mould during sheet forming. A traditional deckle overlaps the sides of the mould, while a basic flat deckle is held in place by your hands. The deckle determines the shape and thickness of the sheet of paper by preventing the pulp from spilling off the edges of the mould. You can use a mould without a deckle, but the pulp will run off the edges of the mould, resulting in a less well formed sheet of paper. Pulp which slips under the deckle during sheet forming creates the characteristic, irregular 'deckle' edge of handmade paper.

There are two types of European moulds: 'laid' and'wove'. Both terms denote the type of screen surface attached to the mould frame. A wove mould refers to any mould with a woven mesh surface. The covering cloth can be made out of woven fabric, finely woven phosphor bronze wire or polyester screening, and is designed to leave little imprint on the paper. Paper made on a wove mould is called wove paper. The surface of a European laid mould is composed of closely spaced, horizontal (laid) wires, which are literally laid across the frame, and held together with finer more widely spaced perpendicular (chain) wires. Paper made on a laid mould is called laid paper - a lined impression, resulting from the laid and chain wires, can be discerned on most laid papers. The screen surface of European moulds is permanently attached to the mould frame. On larger moulds especially, wooden ribs running parallel to the shorter edges of the mould and an extra layer of screen material provide the strength needed to withstand the extra weight involved in scooping up larger amounts of water and pulp from the vat.

Japanese papermaking moulds are called *sugeta*. The *su* is made of slender, closely spaced parallel strips of bamboo, which are held together with fine silk threads. It resembles a European laid mould cover in this respect, but, while the screen surface of a European mould is fixed rigidly to the frame, the Japanese *su* is flexible and is removed from the mould frame once the paper has been formed. The *geta* consists of two lightweight wooden frames, which are hinged along the furthest edge. Once the upper and lower frames are closed in position, the *su* is held securely between them. When the sheet of paper has been formed, the uppermost frame of the *geta* is lifted up and the *su* is removed.

A third type of mould, a wooden frame with a woven cloth stretched across the bottom, is characteristic of the long tradition of Himalayan and Southeast Asian papermaking. The sheet is formed by pouring the pulp into the mould as it floats in a shallow pool of water. Variations on this type of mould are common.

NOTE Contact paper mills for secondhand moulds. When searching for home-made frames and meshes, screenprinting suppliers can often help with enquiries especially for coarse meshes and adhesives. (See also PAPER KITS and ORIENTAL PAPERMAKING.)

ARGENTINA

■ ALEJANDRO D. GEILER
Santa Cruz 1541
8332 Gral Roca
Rio Negra

Mouldmaker
Moulds made from selected woods (cedar, walnut, Brazilian pine, rauli) with bronze wire. Japanese-style also. All sizes and materials available.

■ EL MOLINO DEL MANZO
Fondo de la Legua 375
1642 San Isidro
Buenos Aires
Tel: (54)1 763 8682

Moulds and Deckles
Made out of marine plywood and covered with polypropylene mesh. A4, A5, A6 sizes kept in stock; other sizes to order.

AUSTRALIA

■ GERALDTON PAPERMAKERS
PO Box 145
Geraldton, WA 6530

Moulds and Deckles
Quality-made, pine moulds and deckles. Standard and custom made sizes are available.

■ FRED NICHOLLS
1139 Waterworks Road
The Gap, Qld 4061
Tel: (07) 300 4374

Moulds and Deckles

■ OYSTER BAY PAPER CRAFTS
Roderick & Margaret Graham
4 Tiranna Place
Oyster Bay, NSW 2225
Tel: (02) 528 5008

Moulds and Deckles
From A6 to A2 (single and multiple sheet). Other sizes to order. Dedicated envelope moulds and deckles (not overlays) to Australian post-preferred sizes, as well as individual designs. All frames are polyurethane-finished hardwood, peg-located with stainless steel mesh only. Various mesh sizes available. The larger moulds and deckles have double mesh and ribs, with the support mesh sewn to the ribs.

■ PAPER CAPERS
De-Arne King
PO Box 281
281 Neutral Bay, NSW 2089
Tel: (02) 964 9471

Moulds and Deckles
A5, A5 envelope and A4 moulds and deckles kept in stock. Other sizes made to order. Frames made from Australian timbers such as Radiata pine and cedar; Fibreglass or heat-shrinking, polypropylene meshes.

■ THE PAPER MERCHANT
316 Rokeby Road
Subiaco, WA 6008
Tel: (09) 381 6489
Fax: (09) 381 3193

Moulds and Deckles
From A5 to A3 and larger on request. Envelope, postcard and business card size also available; Japanese *sugetas*. (See ORIENTAL PAPERMAKING.)

Synthetic Felts

■ PRIMROSE PAPERWORKS
PO Box 152
Cremorne, NSW 2090
Tel: (02) 909 1277

Moulds and Deckles

Quality-made moulds and deckles crafted in Western Red Cedar in traditional design.

■ LOIS & BARRY PROSSER
Xanthorrhoea House
49 Nutbush Avenue
Pleasant Grove, WA 6210
Tel: (09) 582 2452

Moulds and Deckles
Made to order from 'environmentally friendly' materials.

■ STEAM FILM PTY LTD.
Peter Zerbe
5 Dunsterville Street
Sandringham
Victoria 3191
Tel: (03) 598 6076

Moulds and Deckles
1 A4 size sold as a set including mould, deckle and pattern deckle of your choice.
2 A4 size separately including mould, plain deckle; envelope deckle, divided deckle, replace wire mesh on mould.
3 A3 moulds and deckles.
4 A2 moulds and deckles.
Other sizes made to order.

CANADA

■ KAKALI HANDMADE PAPER INC.
1249 Cartwright Street
Granville Island
Vancouver, B.C. V6H 3R7
Tel: (604) 682 5274

Custom-made Moulds and Deckles

Student Quality Moulds and Deckles
Polypropylene screening. Finished paper size is approximately 1 4 x 5.5in, 2 7 x 9in.

Synthetic Felts
Foam felts, easy to use, quick drying and machine washable. 1/8in thick, 66in wide, cut to size. Also sold precut to fit 5 x 7in mould.

Wool Felts
100% wool felt, white, 13mm thick, 1.5m wide. Seconds sometimes available. Greater widths available by special order.

■ THE PAPERTRAIL
1546 Chatelain Avenue
Ottawa, Ont K1Z 8B5
Tel: (613) 728 4669
Fax: (613) 728 7796
Orders: 1 800 363 9735

Canadian-made Wove Moulds and Deckles
Both cut from fibreboard, waterproofed and covered with heat-shrinking polypropylene screening. The deckles fit over the side of the mould to prevent slippage. Two models:
1 Basic Moulds and Deckles a. 5 x 7in, b. 8.5 x 11in, c. 11 x 14in
2 Ribbed Moulds and Deckles a. 8.5 x 11in, b. 11 x 14in, c. 12 x 18in, d. 16 x 21in, e. 20 x 26in, f. 22 x 30in

Custom Moulds and Deckles
Laid moulds and deckles in all sizes, made from mahogany with copper laid and screen wires, copper strips around the top of the mould and brass strips over the corners for heavy-duty performance. Made by hand in the centuries-old European tradition.

Custom-shaped Deckles to fit any size mould

Envelope-shaped Deckles to fit various mould sizes

Bathtub Moulds and Deckles
A ribbed, wove, bathtub mould and deckle; fits all standard size North American bathtubs.

Synthetic Felts
Synthetic couching felts up to 5.5m wide. The same felts are used in commercial paper mills. Individual pieces cut to size. Price per square metre.

Wool/Rayon Felts
White wool/rayon felt, 3mm thick, 900mm wide; also suitable for drying paper. Price per metre and individual pieces cut to size.

Polyester Fabric
Non-woven, interfacing-type felt. A cost-effective substitute for real felts and suitable for use on a vacuum table and in pulp spraying. Available in three weights.

Dryer Felts
A highly porous, relatively rigid, synthetic sheeting used in papermills. Can also be used to create airspace in paper dryers. Price by the square metre and individual pieces cut to size.

DENMARK

■ AV-FORM
Hammershusvej 14c
DK-7400 Herning
Tel: 97 22 22 33

Moulds and Deckles
In A4 and A5 sizes, with metal frame and polyester screening.

- HELMUTH OPPENHEJM
  Lojeltegard 16-18
  DK-2970 Horsholm
  Tel: (42) 18 01 22

Wool Felts
TD9, 2mm thickness, 1800mm wide. Suitable for papermaking.

GERMANY

- EIFELTOR MÜHLE
  Auf dem Essig 3
  D-53359 Rheinbach-Hilberath
  Tel: (22 26) 2102
  Fax: (22 26) 2102

Moulds and Deckles
1 In sizes A3, A4, A5 sizes. Made in ash, these have ribs and two layers of bronze screening to ensure better drainage.
2 In A3, A4, A5 sizes. Aluminium with nylon mesh.

Cotton felts
Extremely absorbent, easily-cleaned, cotton felts with an unstructured surface.

- DRUKERN UND LERNEN
  Bleicherstrasse 12
  D-2900 Oldenberg
  Tel: (0441) 1 63 34
  Fax: (0441) 1 40 88

Moulds and Deckles in sizes A3, A4. Contact Peter Voge.

INDIA

- PORRITT AND SPENCER
  ASIA LTD.
  PO Box 384
  Kanchenjunga Building
  No 308/309
  No 18 Barakhamba Road
  New Delhi 110001

Wool-content felts (part synthetic) to traditional specifications

SOUTH AFRICA

- LE PAPIER DU PORT
  Victoria Wharf
  The Waterfront
  Cape Town 8001
  Tel: (0027) 21212305

Moulds and Deckles

SWEDEN

- SANNY HOLM
  Handpapperbruket
  Torsgatan 31
  S-11321 Stockholm

Moulds and Deckles
Both wove and laid moulds, made in the traditional manner. in A3 and A4 sizes. Any other size to order.

SWITZERLAND

■ BASLE PAPER MILL
St Albans-Tal 37
CH-4052 Basle
Tel: (61) 272 96 52
Fax: (61) 272 09 93

Synthetic Felts
Cut to sizes A3, A4, A5 and 25 x 35in.

■ PAPIERATELIER
Teufenerstrasse 75
CH-9000 St Gallen
Tel: (071) 23 50 66

Moulds and Deckles
Various sizes and shapes. Contact Susanne Zehnder or H. Böckle.

UK

■ EDWARD AMIES & CO. LTD.
33 Amsbury Road
Coxheath, Nr Maidstone
Kent ME17 4DP
Tel/Fax: (01622) 745758

A company, founded in 1793 by Edwin Amies, that have made moulds in this area ever since. Formerly based at Hayle Mill, established and traditional makers of traditional laid and wove moulds and deckles. Custom wove and laid moulds and deckles. Also watermarks.

■ BARCHAM GREEN &
COMPANY LIMITED
Hayle Mill
Maidstone, Kent
ME15 6XQ
Tel: (01622) 692266
Fax: (01622) 756381

Historic Moulds
Suitable for both papermaking, display, museums, teaching, etc. dating from 1891-1986. Send two international reply coupons for complete list.

Wool Felts
Good, second-hand (and some new), 100% wool felts made to highest, traditional, professional standards.

■ P&S TEXTILES LTD.
Hornby Street
Bury, Lancs BL9 5BI
Tel: (0161) 764 8617

Wool Felts

Coucher, pron. cōōcher

■ T. McMILLAN
59 Crown Street
Harrow on the Hill
Middsx HA2 OHX

Mould and Deckles
Wove and laid moulds and deckles made to specifications

USA

■ CARRIAGE HOUSE PAPER
79 Guernsey Street
Brooklyn, New York 11222
Tel/Fax (718) 599 PULP (7857)
Orders: 1 800 669 8781

Basic Mould and Deckles
All-purpose moulds made from Baltic birch plywood with a wove surface of poly screening. Range of three sizes.

Envelope Deckles
A deckle that fits over the mould in place of the regular deckle in two sizes.

Dowel Ribbed Moulds
These moulds are more suited to constant use and paper production and are available in larger sizes. Constructed from mahogany with finger joints in the corners. Ribs are made from 3/8in aluminium tubing. Screen surface consists of two layers of poly mesh glued to the frame. Four sizes available.

Polyester Fabric Felts
A non-woven material similar to interfacing fabric. Also excellent as a support material in vacuum table work and pulp spraying. Available in three weights, sold by the linear yard.

■ GOLD'S ARTWORKS, INC.
2100 North Pine Street
Lumberton, NC 28358
Tel: (919) 739 9605
Orders: 1 800 356 2306

Moulds and Deckles
Custom-made moulds and deckles made from 0.75in poplar moulding with a twill weave polyester screen attached with rust-free staples. Various sizes available.

Synthetic Felts
Off-white approximately 3/16in thick. A good substitute for traditional wool papermaking felts. Sold by linear yard which is 36in wide.

■ INTER-OCEAN CURIOSITY STUDIO
Raymond D. Tomasso
2998 South Bannock Street
Englewood, CO 80110
Tel: (303) 789 0282

Used papermaking felts manufactured by Appleton Mills.

■ LEE S. McDONALD, INC.
PO Box 264
Charlestown, MA 02129
Tel: (617) 242 2505
Fax: (617) 242 8825

Basic and Student Moulds
These are the most popular of this company's range and are suitable for beginners. Both are routed out of Baltic Birch plywood with Blue Weave, heat-shrinking, polyester screen. Complete with deckle. Available in two sizes.

Big Basic Moulds
Similar construction to the above makes 11 x 14in paper. Also Big Simple Mould making 11 x 17in paper.

Simple Moulds
Made out of kiln-dried, white oak with a fitted Honduras mahogany deckle. Finger joints for extra strength, and the underside of the deckle has been routed to assure aligment to the mould. Makes 8.5 x 11in paper. The Envelope, Circle and Oval deckles will fit this mould.

Buttercut for Shaped Deckles
Easy to cut, suprisingly strong and easy to make detailed shapes for stencils. With adhesive backing, it is placed directly on the mould. Two or more layers can be stuck together and used as a deckle.

Apprentice Moulds
Finely-constructed, mahogany stock with finger joints and smooth, rounded, bottom edges. Ribs are 3/8in aluminium tubing with two layers of polyester mesh stretched and glued to edges. The deckles have tenon joints for maximum strength. Sizes up to 16 x 20in.

Dowel Ribbed Moulds
An intermediate-duty mould for frequent but not heavy daily use. Sewn construction with backing of polyester screen. The ribs are aluminium dowels. Woven polyester or brass screen can be used as top surface, in sizes up to 18 x 24in.

Custom Moulds
Carefully crafted from straight-grained Honduras mahogany with clear Ponderosa pine ribs. Dovetailed corners are fastened with wooden pegs and high-strength waterproof glue. After assembly, the ribs of the mould are hand-shaped to a slight, convex curve to ensure tight deckle fit and proper sheet formation. Brass strips and escutcheon pins complete a fine making. Various types are made-to-order from Woven to Antique Laids in many different sizes.

CC Felts/Premium Wool Felts
100% woven wool felts. Natural ivory colour. It is necessary to add 10% to ordering size to allow for shrinkage. Sold by linear foor. Various widths. Available in 2 weights.

17L Felt
A good-quality felt made out of 100% polyester. Range of sizes.

60/40 Felt
Fine, cream, couching felt made of blend of 60% wool and 40% polyester. Add 10% for shrinkage when ordering. Sold by linear foot or cut to size. 90in wide.

Ultra Felt 11
A strong, very thin, white, non-woven 100% polyester material. Used to support wet paper on a vacuum table. Not recommended as a couching felt. Sold by linear yard. 45in wide.

■ MAGNOLIA EDITIONS
2527 Magnolia Street
Oakland, CA 94607
Tel: (510) 839 5268
Fax: (510) 893 8334

Various moulds including Basic Mould made from Baltic birch plywood with coarse polyester screen, in two sizes. Also Shaped Deckles made for an envelope size to fit 8.5 x 11in basic mould.

Thick, non-woven, synthetic Interfacing

■ GREG MARKIM INC.
PO Box 13245 Milwaukee
WI 53213
Tel (414) 453 1480 /(800) 453 1485
Fax (414)453 1495

Couch Sheets
In packs of 25's and 100's. (See also PAPERMAKING KITS.)

■ PARAGON PAPER MOLDS
Timothy Moore
14450 Behling Road
Concord, MI 49237
Tel: (517) 524 6318

Laid Moulds, Fine Laid Moulds, Traditional Wove Moulds, Poly-faced, Poly-backed wove Moulds and Deckles. custom-made with 100% phosphor bronze facings, standard (9 laid wires per inch) or fine (26 laid wires per inch) laid backing wire, vertical grain, water-cured wood. Every effort is made to make the finest moulds obtainable. Traditional joinery and waterproof glue, dovetail joints; large moulds have brass corner reinforcements for added rigidity. All wove moulds are double-faced; laid facings are made in house on a wire weaving loom. Stock sizes on request or custom-made. Various special items can be made to order. Contact Timothy Moore.

■ SUBMARINE PAPERWORKS
Jim Meilander
PO Box 885082
San Francisco, CA 94188-5082
Tel: (512) 224 1848

Deckle Boxes
Make sheets of paper without using a vat with a Vietnamese-style, deckle box. Basic 8.5 x 11in, solid mahogany, brass hardware, polyester screen. Wide range of other sizes.

■ TWINROCKER HANDMADE PAPER
PO Box 413
Brookston, IN 47923
Tel: (317) 563 3119
Fax: (317) 563 8946

Ribless Moulds
A sturdy mould with polyester surface stretched so tightly that ribs are not required. 2 sizes.

Deckles
The larger of the two moulds (8.5 x 11in) is made with either a traditional deckle (overlap) or a basic flat deckle that is held in place with your hands.

Shaped Deckles
Constructed of Baltic Birch plywood. In three shapes - Envelope, Oval and Circle Deckles.

Dishpan Moulds

Wool Felts

Mouldmaking components

CANADA

■ KAKALI HANDMADE PAPER INC.
1249 Cartwright Street
Granville Island
Vancouver, BC V6H 3R7
Tel: (604) 682 5274

Polypropylene Screening
Sold by the square foot to a maximum 6ft wide.

■ THE PAPERTRAIL
1546 Chatelain Avenue
Ottawa, Ont K1Z 8B5
Tel: (613) 728 4669
Fax: (613) 728 7796
Orders: 1 800 363 9735

Screening
White, papermill screening available in varying widths from 1 to 3 metres. Priced by the square metre. Individual pieces cut to size.

Heat-Shrinking Polypropylene Screening
Easy to retighten across the mould through the application of heat.

USA

■ CARRIAGE HOUSE PAPER
79 Guernsey Street
Brooklyn, New York 11222
Tel/Fax (718) 599 PULP
Orders: 1 800 669 8781

Heat-Shrinking Polypropylene Screening

Monel stainless staples by Arrow; Arrow T-50 manual Staple Guns. Electric Staple Guns

■ GOLD'S ARTWORKS, INC.
2100 North Pine Street
Lumberton, NC 28358
Tel: (919) 739 9605
Orders: 1 800 356 2306

Screening
A twill-weave, heat-shrinking polyester screen used for Gold's Custom Moulds.

■ LEE S. McDONALD, INC.
PO Box 264
Charlestown, MA 02129
Tel: (617) 242-2505
Fax: (617) 242-8825

Flat Brasses and Corners: Various flat brass trims for edging the screen and corners for traditional mould making. Also Brass Escutcheon Pins and Brass Half-round Corners.

Blue Weave: Light blue, heat-shrinking screen polyproplene screen used on Basic and Student moulds; Epoxy Paste: two-part formula for glueing screens onto moulds.

6 x 6 Polyester: Lightweight, white backing especially useful for moulds where weight is crucial. Used in combination with brass or polyester top surfaces. 56in wide.

Meshes
A number of brass meshes for top and backing. Also Phosphor bronze for laid surface with 1in chain spacing and 19 wires per inch laid lines. Sold cut-to-size only.

Bronze backing
10 x 10 bronze backing used as a backing wire for laid surface, produces paper with a less pronounced laid pattern. Sold cut-to-size only.

- MAGNOLIA EDITIONS
  2527 Magnolia Street
  Oakland, CA 94607
  Tel: (510) 839 5268
  Fax: (510) 893 8334

Screening
A heat-shrinking polyester screen used on Magnolia's Student Moulds.

- TWINROCKER HANDMADE PAPER
  PO BOx 413
  Brookston, IN 47923
  Tel: (317) 563 3119
  Fax: (317) 563 8946

Wire Cloth

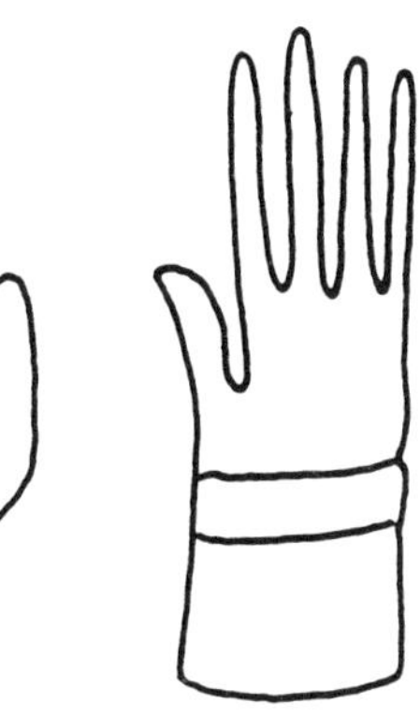

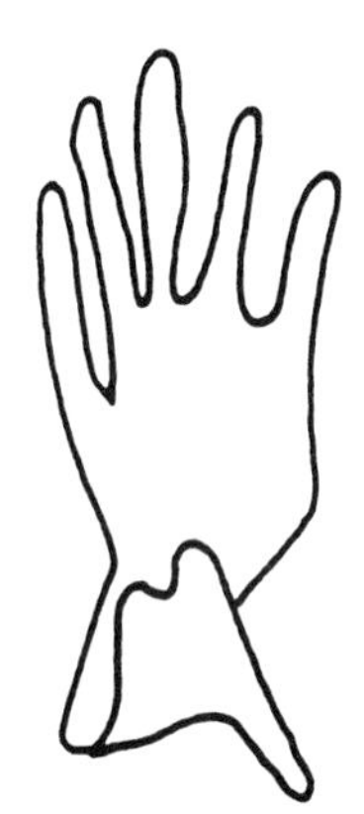

Watermarks

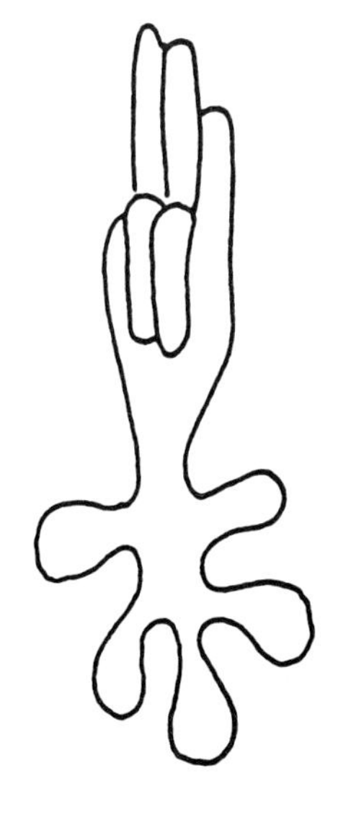

The term 'watermark' is used to describe the impression left in a sheet of paper usually appearing as an initial or an outlined image when the sheet is held up to the light. A watermark is made by attaching a fine wire design to the surface of the mould. As the pulp rests on the surface of the mould, less pulp remains where the wire is thus allowing a thinner layer of paper in the form or image of the wire design.

In the past a great variety of pictorial designs were employed, including animals, humans, plants, heraldic motifs, simples shapes such as crosses, stars, circles, often to identify the mould belonging to illiterate papermakers. Today watermarks serve mainly as the logo or trademark of a particular mill or papermaker.

When a hidden image is expressed in terms of light and shade, as opposed to a simple outline, it is called a 'chiaroscuro' watermark. It is made by taking a plaster cast from an image that has been carved as a bas relief into a wax block. A die and matrix are produced from the cast and impressed into a wire cloth of a wove screen before it is attached to the mould. When the sheet of paper is formed thinner and thicker areas of pulp are deposited in the undulating relief areas of the mould's surface. When the sheet has been pressed and dried, the tonal watermark can be clearly discerned by holding the sheet up to the light.

SWITZERLAND

■ BASLE PAPER MILL
Swiss Paper Museum and
Museum for Script and
Printing
St. Alban-Tal 35/37
CH-4052 Basle
Tel: (61) 272 96 52
Fax: (61) 272 09 93

Write to the mill for details.

Watermark device on a wove mould made by Edward Amies & Co. for Barcham Green

UK

■ EDWARD AMIES & CO. LTD.
33 Amsbury Road
Coxheath, Nr Maidstone
Kent ME17 4DP
Tel/Fax: (01622) 745758

Contact Mr. R. J. MacDonald. A company founded in 1793 by Edwin Amies. Formerly based at Hayle Mill. Established makers of traditional laid and wove moulds and deckles and watermarks.

USA

■ LEE S. McDONALD, INC.
PO Box 264
Charlestown, MA 02129
Tel: (617) 242 2505
Fax: (617) 242 8825

Almost any design as well as signatures can be easily reproduced with varying line widths. Lines should be at least 1/8in apart for clarity. Black and white camera-ready art work (no line drawings). Watermarks can be sewn to your mould also. Photo-etched magnesium watermarks are available and traditional bent wire marks can be made (quote only basis).

■ SEA PENN PRESS &
PAPERMILL
2228 N.E. 46th Street
Seattle, WA 98105
Tel: (206) 522 3879

Wire watermarks and shadowmarks can be commissioned. The artist/designer submits either a simple line drawing for translation into wire or a matrix for a shadowmark such as a sculpted relief in wood, plaster, or a collograph plate. The shadowmark differs from a traditional linear watermark in that a copper mesh is shaped into a bas-relief an 1/8in or less in height using repousse techniques or a press to push the wire mesh into the matrix.

Vats

A nineteenth century vat

The vat or tub which contains the pulp and water for papermaking has changed little over time. Various devices were developed to assist the sheet maker, for example a heating device called a 'pistolet', a 'horn' (flat plank across the back of the vat supporting a notched upright against which the mould was placed to drain prior to couching); mechanisms such as a 'knotter' (which filtered the knots and other impurities from the pulp before it entered the vat), and a 'hog' (a rotating paddle at the bottom of the vat which kept the pulp in suspension), were added to later production. Vats have been made of stone or iron and lined with lead to prevent rust. Contemporary production vats for European and Japanese papermaking usually follow a rectangular design, with the front wall slanting in towards the base. Wood (pine and cedar), stainless steel sheeting, concrete lined with ceramic tiles and marine fibreglass are usual. Vats made of polyethylene provide a light-weight and durable alternative. When choosing a suitable vat you will need to consider the size of your mould and deckle, allowing enough depth to scoop up the pulp and a comfortable clearance between your mould as you hold it and the sides of the vat. Perhaps most importantly, the height of the vat in relation to your own size must not be neglected, in order to avoid placing undue strain on the muscles in your back as you lift the mould up out of the vat.

CANADA

■ KAKALI HANDMADE PAPERS, INC.
1249 Cartwright Street
Vancouver, B.C. V6H 3R7
Tel: (604) 682 5274

Black or grey Plastic Vats, size 13.5 x 18.5 x 7in.

■ THE PAPERTRAIL
1546 Chatelain Avenue
Ottawa, Ont K1Z 8B5
Tel: (613) 728 4669
Fax: (613) 728 7796
Orders: 1 800 363 9735

Plastic Vats with lids approx. 35 x 48 x 22cm deep. Black Polyethylene Vat (without lid): (Small = 66 x 50 x 5cm deep; Large = 90 x 60 x 20cm). Plastic Laundry Tub Vats with detachable legs. Rectangular White Polyethylene Tanks (with drain) which are designed for making large sheets and needs to be mounted on a platform and braced to prevent bulging; size 132 x 107 x 61cm.

USA

■ LEE S. McDONALD, INC.
PO Box 264
Charlestown, MA 02129
Tel: (617) 242 2505
Fax: (617) 242 8825

Lightweight, durable vats made of black polyethylene and drop-shipped, larger-sized vats. Black Vats: 17in length x 23in width x 6in depth. Also 33 x 21 x 8in. Drop-shipped Vats: White polyethylene, tapered at bottom with a special lip for improved strength. Range of sizes.

■ MAGNOLIA EDITIONS
2527 Magnolia Street
Oakland, CA 9460
Tel: (510) 839 5268
Fax: (510) 893 8334

Small Vats: 24 x 16 x 9in deep; Large Vats: 35 x 24 x 23in deep.

Until about 1950 we relied entirely on sunshine for drying wet *washi*. . .

Top Paper drying in a terraced field, Nepal
Bottom Maureen Richardson of Plant Papers operating her press surrounded by bags of plant material

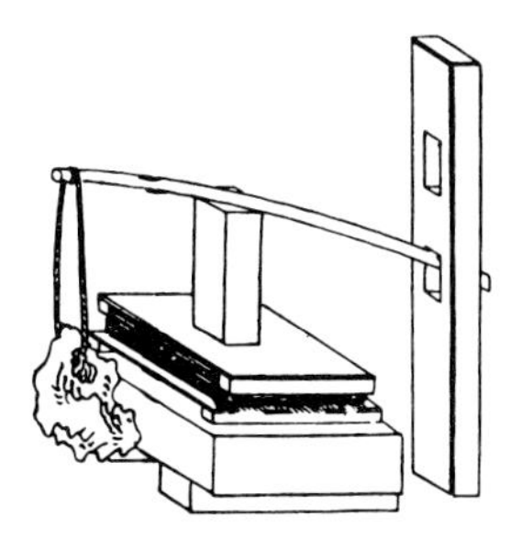

Paper presses
Paper dryers

can still vividly picture the scene, in early, Spring, when heet after sheet of paper was sucked up into the air by a udden gust of wind and blown far into the distance.

(Top and bottom quotes) Minoru Fujimori in Claire Bolton's *Awa Gami* 1992

Paper presses

Hydraulic Press by Oyster Bay Paper Crafts

A newly made sheet of paper contains a great deal of water (roughly 90%). The function of a press is to extract as much of this water as possible and to help bond the fibres into a strong sheet, by exerting an even surface pressure on top of the paper. In traditional Western hand papermaking, massive wooden screw presses were once used to force the water out of a post of recently couched paper. The sheets were sometimes parted again from the felts and given a second, longer pressing between fresh dry felts using less pressure. They were often re-stacked without interleaving felt, and pressed again several more times in order to diminish the mark of the felts and create a smoother surface. Today, powerful hydraulic presses have largely replaced their earlier counterparts.

Early Eastern pressing methods consisted of several large stones placed on a pressing board on top of the post of paper, or a simple lever press hung with incremental weights. The traditional Japanese method of pressing is far more gradual than the equivalent Western process. The post is usually covered with a cloth and left to drain overnight. It is then pressed slowly for several hours using a light pressure, which is only increased a little at a time.

Custom-designed drying systems, based on the layering system botanists use for drying plants, provide a practical means of keeping paper flat during the drying process. Damp sheets of paper, or low relief artwork, can be placed between layers of archival double-wall corrugated board or mesh screen. A platen is used to apply pressure on top of the stack, and a fan blower draws out the moisture.

AUSTRALIA

- FRED NICHOLLS
  1139 Waterworks Road
  The Gap, 4061 Brisbane
  Queensland, W. Australia
  Tel: (07) 300 4785

  Paper Presses

- OYSTER BAY PAPER CRAFTS
  4 Tiranna Place
  Oyster Bay, NSW 2225
  Tel: (02) 528 5008

  Presses
  Manufacturer of hand papermaking and ancillary equipment. Hydraulic presses to A2, screw presses to A4. Contact Rod Graham.

- THE PAPER MERCHANT
  316 Rokeby Road
  Subiaco, WA 6008
  Tel: (09) 381 6489
  Fax: (09) 381 3193

  Paper Presses

- LOIS & BARRY PROSSER
  Xanthorrhoea House
  49 Nutbush Avenue
  Peasant Grove 6210
  Western Australia

Simple Presses
Basic presses with a screw thread on each corner, for A3.

Advanced Presses
Steel frame plus a single, central screw thread, for A3 and A4.

Pressing Boards

- STEAM FILM PTY LTD.
  Peter Zerbe
  5 Dunsterville Street
  Sandringham
  Victoria 3191
  Tel: (03) 598 6076

Hydraulic Presses: 20 ton press.

CANADA

- THE PAPERTRAIL
  1546 Chatelain Avenue
  Ottawa, Ont. K1Z 8B5
  Tel: (613) 728 4669
  Fax: (613) 728 7796
  Orders: 1 800 363 9735

Screw Presses
This is a wooden, manual screw press, fully urethaned for waterproofing. The platens measure 24 x 30in and are designed like butcher's blocks for proper distribution of forces. It stands on the floor, has wheels and a tray to hold utensils. Smaller presses of same type are also available.

GERMANY

- EIFELTOR MÜHLE
  Claudia Stroh Gerard
  Auf dem Essig 3
  D-53359 Rheinbach-Hilberath
  Tel/Fax (22 26) 2102

Presses
Sturdy metal presses which includes base tray to collect run-off water. Can be used for up to 5 tons with standard hydraulic car jack.

INDIA

- SITSON INDIA
  W 76, M.I.D.C. Phase II
  Dombivli (East) 421 204
  District Thane
  Maharashtra State
  Tel: (251) 666553
  Fax: (251) 466950

Hydraulic Presses
Heavy-duty, up-stroking, hydraulic presses complete with pumps and controls. Up to 100 ton force, possibly more. Sitson also make production Hollander beaters and other equipment.

NETHERLANDS

- PETER GENTENAAR
  Churchilllaan 1009
  NL-2286 AD Rijswick
  Tel: (1742) 6961

Paper Presses
Basic, screw/spindle presses, varying in sizes from 40 x 60cm to 80 x 80cm and larger.

USA

- CARRIAGE HOUSE PAPER
  79 Guernsey Street
  Brooklyn, New York 11222
  Tel/Fax: (718) 599 PULP (7857)
  Orders: 1 800 669 8781

Aardvark Presses
A medium-sized, versatile press, light enough to make it portable. The frame is steel coated with rust-proof enamel. 12 ton hydraulic jack and pressure gauge gives the user greater control. Drain tray and two boards, size 16 x 23in.

Reina Hydraulic Presses
A heavy-duty, aluminium press that comfortably allows pressing of sheets up to 24in wide. Includes 25 ton hydraulic jack plus a pressure gauge.

Custom Presses (built to any size)

- GREG MARKIM, INC.
  PO Box 13245
  Milwaukee, WI 53213
  Tel: (414) 453 1480
  Fax: (414) 453 1495

Professional Hydraulic Jack Press Frame and Blocks
An appropriate-sized press frame designed around an hydraulic car jack (purchased separately).

- LEE S. McDONALD, INC.
  PO Box 264
  Charlestown, MA 02129
  Tel: (617) 242 2505
  Fax: (617) 242 8825

Howard Clark 20 Ton Presses
Originally designed by Howard Clark of Twinrocker Handmade Paper. Manufactured exclusively by Lee S. McDonald, Inc., this reliable press is made of welded, structural aluminium. With a 20 ton hydraulic jack, the Howard Clark 20 Ton Press is capable of pressing sheets of paper up to 22 x 30in in size.

The Kirby Portable Presses
A finely-crafted, wooden press made from maple/Baltic Birch including a 6 ton jack, special steel jack handle and 18 x 18in plywood platens. The Kirby Press is easily dismantled for transportation and storage. Dimensions are 22.25in wide x 45in high x 18in long.

Bookbinder's Presses
A bookbinder's model of the Howard Clark 20 Ton Press but with a wider opening of 34in (instead of 25in) for glueing books. Presses 30 x 40in paper (one half at a time).

■ TWINROCKER HANDMADE PAPER
PO Box 413
Brookston, IN 47923
Tel: (317) 563 3119
Fax: (317) 563 8946

Arnold Grummer Hydraulic Jack Presses
These consists of a sturdy press frame and press boards for either 6 x 8in or 8.5 x 11in paper and are designed for use with a 2 ton hydraulic car jack. You may specify larger press boards at an extra cost.

Paper dryers

It is important that as the paper dries out, a uniform moisture content is obtained for each sheet and that the water is propelled outwards from the centre of the pile in the pressing operation. Often up to 60-70% of the water is expressed by running the sheets through a system of pressure rollers and it is then laid or hung to complete the drying process. History dictates that the final drying of the handmade sheets takes place in a special drying area. Many mills in Europe were built with a special drying loft in which the paper was laid or hung high up at the top of a building where the air was supposed to be cleaner and fresher and where a prevailing breeze could be caught, avoiding the soot, dust and dirt from lower floors. Today many mills employ heated drying operations where the sheets are passed through a heated cabinet or over hot cylinders.

CANADA

■ THE PAPERTRAIL
1546 Chatelain Avenue
Ottawa, Ont. K1Z 8B5
Tel: (613) 728 4669
Fax: (613) 728-7796
1 800 363 9735

Paper Dryers
Front-loading paper dryer with inside measurements 70 x 90cm; equipped with a large fan that pulls air through the drying stack to speed up the drying process.

PHILIPPINES

■ BERNARD HANDMADE PAPER PRODUCTS
Concepcion, San Pablo City
Phillipines 4000

Dryers
For efficient drying of handmade sheets especially during the rainy season. Made of black iron sheet on black iron angle-bar frame. Size 18 x 24in.

Screw Presses
For processing of sized papers. Made of heavy steel channels and slated with centrally-located, screw/nut/gear train assembly. Receiving plate is 24 x 36in taking 40 sheets of paper.

Calender Machines
This machine is used to produce smooth-surface papers and to attain uniformity in paper thickness and includes a pair of steel rollers mounted on a platform and driven by a series of gear/pulleys.

USA

■ LEE S. McDONALD, INC.
PO Box 264
Charlestown, MA 02129
Tel: (617) 242 2505
Fax: (617) 242 8825

The Santa Ana Paper Dryer
Designed to keep paper flat during the drying process. The Santa Ana Paper Dryer provides a useful facility for stacking wet or newly made paper in between layers of packing materials made especially for flat drying. The air is pulled through the stack via a blower in order to dry the paper. Constructed with Baltic birch plywood, the Santa Ana Paper Dryer consists of a platen and blower plenum which creates greater air pressure within the enclosed space than that of the outside atmosphere. Strap winches on the top pull down for pressure to keep your work flat; the platens act as a convenient way to apply pressure. The unit is easy to load and unload by removing the top platen and separating the paper from the packing material. The standard stack of packing materials is sold separately from the drying machine. The standard Santa Ana Paper Dryer size is 36in wide x 48in deep x 44in high and can be loaded with 10 or 20 layers of packing material and art work. Comes with locking casters for mobility. The following options are also available: bottom mounting of the blower which makes a more compact unit as the blower is mounted on the underside. A tilted table top can be added for a work surface where space is limited.

Packing Materials
The standard packing material for the stack is the Screen Sandwich. Each Screen Sandwich layer consists of a core of diamond mesh screen and two pieces of aluminium screen mesh with two polypropylene felts on the outside, which, when assembled, create an air space for moisture removal. Versatile and long lasting, the Screen Sandwich will prevent most papers from cockling when proper pressure is applied. The packing material is cut to size for the Santa Ana Paper Dryer.

Custom Dryer Systems
Custom-designed systems for drying art work or flat sheets.

Electric Drying Boxes
A welded frame with castors and a screw press restraining system. 30 x 40in capacity; blotters included. Capable of drying approximately one hundred 30 x 40in sheets. Custom sizes are available.

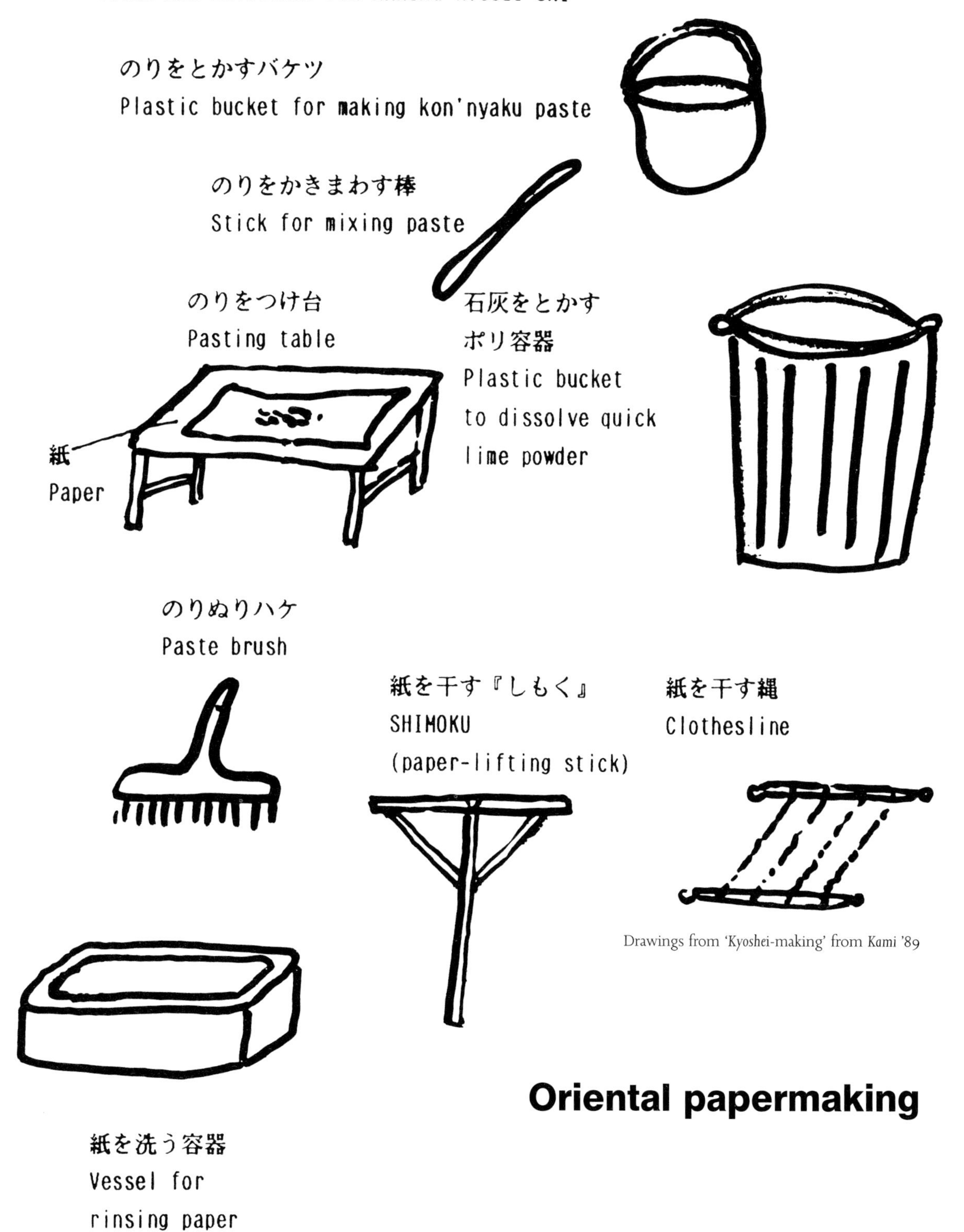

Drawings from 'Kyoshei-making' from *Kami* '89

Oriental papermaking

To learn, on occasions I had to enquire, at other times I was shown without asking, and at still other moments I learned by osmosis, by watching and listening alone.

Timothy Barrett *Japanese Papermaking – Traditions, Tools and Techniques* 1983

Oriental papermaking

Many papermakers are explorers who search out different, lost, unknown or undocumented papermaking practices in countries around the world. What is so fascinating about papermaking is how the processes in each country differ; at each papermaking stage the practice is affected by climate, available vegetation, cultural and aesthetic requirements, etc. We have located a number of suppliers of Japanese papermaking requisites but firstly we describe the process and how it differs from the European method previously described.

Long bast fibres are predominantly used in Japan, compared to the shorter cotton and linen rag fibres of European papermakers. *Kozo, Gampi* and *Mitsumata* are the three plants which have proven most suitable. A special steaming treatment called *seiromushi* softens the outer bark which is stripped off and hung to dry and subsequently stored for future papermaking. The fibres are cooked in an alkali solution for roughly two to three hours depending on the kind of paper being made. They are then rinsed thoroughly in running water and any remaining foreign particles are removed by hand. Beating is a shorter process than in Western papermaking, because long fibre length and slow drainage rate on the mould are essential to good sheet forming. Traditional wooden hand beating paddles are still used, together with bast fibre stampers and a Japanese *Naginata* beater (like a European Hollander, but fitted with curved knives which hack at the fibres and circulate them around the tub). A clear, glue-like extract called *neri* is added to the papermaking solution. It is obtained from the crushed roots of the *tororo-aoi* plant (Hibiscus manihot, Medikus). *Neri* increases the viscosity of the water and slows the drainage of the vat mixture through the screen during the sheetmaking. It also acts as a formation aid by preventing the long bast fibres from becoming entangled during the papermaking process.

Making a sheet

The Japanese papermaker uses a flexible bamboo or reed screen, called a *su*, which is held between a ribbed mould frame and hinged deckle, the *geta*. In this process, the papermaker scoops up a small amount of pulp from the vat and distributes it over the screen. Any excess is thrown off back into the vat. This process is repeated several times, gradually accumulating layer upon layer until the desired thickness in achieved (*Nagashizuki*). It is quite unlike the European method of dipping the mould into the pulp a single time (*Tamezuki*). The *su* is lifted off the mould frame to couch the sheet. The presence of *neri* obviates the need for interleaving felts and the sheets are couched directly on top of each other.

Couching of *washi* from a two-sheet *su* directly on top of a pile of wet sheets

The method of pressing sheets of paper in Japan is far more gradual than the equivalent European process. The post is usually covered and left to weep overnight, before being pressed slowly for several hours under an incremental weight. Oriental papers are dried in various ways. In Japan, drying often takes place outdoors, with papers brushed onto seasoned wooden boards. The side of the paper dried against the board will therefore be smoother than the other. Artificial drying on a heated metal surface or stone wall is also a widespread, though less traditional, practice. Similar drying methods are used in China and, in some papermaking districts, bamboo paper is left to dry on the ground in the sun after pressing.

Pasting wet *washi* onto drying boards

Traditional papermaking in Nepal, Bhutan and other Himalayan countries differs again from the practice of the Japanese and European papermaker. A sheet of paper is formed by pouring a measured amount of bast fibre pulp into a mould floating in a shallow pool of water. When the fibres have settled, the mould is lifted up and placed on end and left in the sun. The paper is not pressed and the sheet is then peeled off the mould when it has dried. Sheets made in this way have a distinctive cloud-like fibre formation.

In India, where the production of handmade paper was revived in the 1920's as part of Mahatma Gandhi's 'khadi' or village industry movement, sheets are made from a long-fibred, handspun cotton mixed with organic fibres for colour and texture. In some areas, European and Asian influences are evident in the papermaking process, which combine the floating-mould method of sheet forming with the use of a Western-type wove mould and deckle to facilitate couching and pressing.

Oriental fibre descriptions

JAPANESE *KOZO*
(*Broussonetia papyrifera*)

Kozo is the most widely used bast fibre in Japan and comes from the inner bark of the *Kozo* (Paper Mulberry) plant. *Kozo* fibre has three layers: the outer black bark, a secondary layer of green bark, and the inner white layer. The fibres are about 10mm long and very strong. The raw fibre must be cooked and, if necessary, the black bark (*chiri*) and green bark layer must be removed by hand; but it is much easier to beat than Western fibres and can, in fact, be beaten by hand or with a stamper. It makes an exceptionally strong, pale yellow paper that retains its strength even when crumpled and folded.

THAI *KOZO*

A less expensive *Kozo*, grown in Thailand. The plant is identical, botanically, to Japanese *Kozo*, but warmer climatic conditions result in a difference in the fibre's characteristics.

CHINA *KOZO*

A little less expensive than Thai *Kozo*, similar to Japanese *Kozo* in appearance.

MITSUMATA
(*Edgeworthia chrysantha*)

A Japanese fibre, *mitsumata* has a fibre length of about 4mm and produces a pinkish/pale tan paper with a soft, smooth and lustrous surface.

JAPANESE *GAMPI*
(*Wikstroemia diplomorpha*)

Gampi is regarded as the most refined of Japanese papermaking fibres. It makes a strong paper with a lustrous, silky sheen and a warm greenish/yellow natural hue.

PHILIPPINE *GAMPI*

Identical to Japanese *gampi*, but the fibre has different characteristics (somewhat coarser than the Japanese variety) due to climate and growing conditions. Produces a crisp, off-white/tan sheet with a lustrous sheen.

Suddenly, I was so strangely thrilled with this subject.

Sukey Hughes *Washi The World of Japanese Paper* 1978

AUSTRALIA

- **THE PAPER MERCHANT**
  316 Rokeby Road
  Subiaco, WA 6008
  Tel: (09) 381 6489
  Fax: (09) 381 3193

Fibres: *Kozo; Gampi; Mitsumata*

Also *Sugetas*; Formation Aid (See CHEMISTRY; Additives); *Konnyaku* Powder: used to make *konnyaku* starch which is a binder and gives Japanese papers extra wet-strength (applied with a brush).

CANADA

- **KAKALI HANDMADE PAPERS, INC.**
  1249 Cartwright Street
  Vancouver, B.C. V6H 3R7
  Tel: (604) 682 5274

Raw Fibres: Japanese *Kozo*; Thai *Kozo*; China *Kozo*; Philippine *Gampi*

Formation Aid

Rabbit Hair Brushes

- **THE PAPERTRAIL**
  1546 Chatelain Avenue
  Ottawa, Ont. K1Z 8B5
  Tel: (613) 728 4669
  Fax: (613) 728 7796
  Orders: 1 800 363-9735

Raw Fibres: Philippine *Gampi*; Thai *Kozo* (more fibres available).

Hand Beating Sticks

JAPAN

- **Hr. ARIMITSU**
  2-9-34 Kami-machi
  Kochi, Kochi Pref.
  Tel: (0888) 72 6876

Sugetas: Traditional Japanese custom-made papermaking moulds.

- **DIAFLOC CO.**
  Flocculant & Papermaking
  Additives Division
  New Marunouchi Bldg.
  51, Marunouchi 1 chome
  Chiyoda ku, Tokyo 100
  Tel: (03) 3213 7191
  Fax: (03) 3215 1785

Retention Agent (See CHEMISTRY; Additives)

Formation Aid (See CHEMISTRY; Additives)

Contact Mr. Choko Takahashi (Manager, Sales Department)

Even if a person doesn't make paper very skilfully, if he doe it honestly, it will be good paper. You can see it in the shee

Seikichiro Goto, from Sukey Hughes *Washi The World of Japanese Pape*

■ FUJI PAPER MILLS CO-OPERATIVE
The Hall of Awa Japanese Handmade Paper
141 Kawahigashi
Yamakawa-cho, Oe-gun
Tokushima 779-34
Tel/Fax: (0883) 42 6120

Raw Fibres: *Kozo*, black bark: This is Tokushima-grown *Kozo* ready for cooking, often used when making paper containing bits of bark (*chiri-iri*), and when bleached the fibre is used for making *shoji* paper, etc.; *Kozo*, white bark: This Kochi-produced *Kozo* is ready for cooking and is used for unbleached *Kozo* paper; Thai *Kozo*: This is a black bark imported from Thailand and can be substituted for domestically grown *Kozo*; *Mitsumata*: A white bark cultivated in Tokushima and Kochi used by the Ministry of Finance for making Japanese paper currency; *Gampi*: A wild *gampi*, harvested in early summer from the Sanuki Mountains.

Ready-to-use Pulps: *Kozo* (wet pulp): An bleached beaten pulp made from Thai *Kozo*; *Mitsumata*: An unbleached, beaten pulp made from Tokushima-grown *Mitsumata*; *Gampi*: An unbleached, beaten pulp made from Tokushima-grown *gampi*. Various sizes of packs available. Write for details.

Steaming *kozo* fibres

Chemicals: *Ginbiaso* (natural *neri*): This plant grows wild in the marshes at the foot of Mt. Kenzan and produces a very special type of *neri*. It is the only type used in Nashio (Hyogo prefecture) for making *hakuai* paper; *Tororo-aoi* (natural *neri*): specially grown in Tokushima; Formation Aid; *Konnyaku* Powder: used to make *konnyaku* starch, it acts as a binder and gives Japanese papers extra wet strength. It is applied by brush. Internal Sizing: Sizing Liquid A and Sizing Liquid B; CMC Methyl Cellulose; Alum; *Nikawa* (animal glue).

Mitsumata seeds; *Tororo-aoi* seeds

Sugetas: Postcard-size *sugeta* (10 x 15cm); *Hanshi*-size *sugeta* (27.5 x 36.5cm); *Kikuban*-size *sugeta* (64 x 97cm)

■ Hr. KIKUCHI
90 Funyu, Yamagata-machi
Naka-gun, Ibaraki-ken 319-31
Tel: (029) 57 2252

Konnyaku Powder

■ YOSHIROU KINUYAMA
2849 Ochi, Ino-cho
Agawa-gun, Kochi 781-21
Tel: (780) 0888 93 2639

Raw Fibres: Japanese *Kozo*; Japanese *Mitsumata*

Sugetas: Custom made. (State size of mould and thickness of the paper you wish to make.)

SWEDEN

■ ZEN ART PAPER AB
St. Jorgens vag 20
S-422 49 Hisings Backa
Tel: (31) 55 68 55
Fax: (31) 55 56 85

Raw Fibres: Banana, Daphne (Lokta), Jute

Moulds and Deckles: Bamboo (25.5 x 43.5cm); Traditional (20 x 15cm); Traditional (30 x 25cm). Contact Christina Bolling.

UK

■ ATLANTIS EUROPEAN LTD.
146 Brick Lane
London E1 6RU
Tel: (0171) 377 8855
Fax: (0171) 377 8850

Japanese Brushes
A range of brushes imported directly from Tokyo. All brushes have been hand-crafted from the finest materials using traditional methods of manufacture .

■ PAPER PLUS
24 Zetland Road
Chorlton, Manchester 21 2TH
Tel: (0161) 881 0672/445 7273

Japanese Papermaking Educational Resource Packs
Contains instructional booklet, slides, Japanese *sugeta*, selection of Japanese paper samples, photographs, paper mulberry and *tororo-aoi*. Contact Maggie Holland for more details.

USA

■ CARRIAGE HOUSE PAPER
79 Guernsey Street,
Brooklyn, New York 11222
Tel/Fax:(718) 599 PULP (7857)
Orders: 1 800 669 8781

Raw Fibres: Japanese *Kozo*; Chinese *Kozo*; Thai *Kozo*; Japanese *Mitsumata*; Japanese *Gampi*; Philippine *Gampi*. Bast Fibre Sample Kit (contains 1lb of each fibre); Formation Aid

Book of Paper from Bast Fibres: This includes 2 (8 x 12in) sheets of each fibre (a total of 12 sheets in all) made on a *sugeta* and bound in the traditional Japanese manner.

Sugetas made in Taiwan: 13 x 16.5in; 16.5 x 18.5in, 18 x 24in (with handles). These light-weight wooden moulds and deckle frames are hinged and have dovetail joints; the *su* is constructed by hand.

■ MAGNOLIA EDITIONS
2527 Magnolia Street
Oakland, CA 94607
Tel: (510) 839 5268
Fax: (510) 893 8334

Raw Fibres: Thai *Kozo*; Korean *Kozo*; Japanese *Mitsumata*

Sugetas: Small 5.5 x 7.5in; Medium 13.5 x 16.5in; Large 18 x 24in (with handles)

Formation Aid

■ LEE S. McDONALD, INC.
P.O. Box 264
Charlestown, MA 02129
Tel: (617) 242 2505
Fax: (617) 242 8825

Raw Fibres: Japanese *Kozo*; Thai *Kozo*; Japanese *Mitsumata*; Japanese *Gampi*; Philippine *Gampi*

Also Naga Brushes; Hand Beating Sticks (for processing bast fibres); Stampers (for preparing bast fibres); *Nagashizuki* presses and *Nagashizuki* vats with a *Maze* (according to availability); Formation Aid

Naginata Beater
Common in Japanese papermaking, the *Naginata* is used following hand or stamper beating. Lee S. McDonald can quote a price based on the design of the prototype published in *Japanese Papermaking: Traditions, Tools and Techniques* by Timothy Barrett.

Health and safety in papermaking

Papermaking and the environment

Paper testing equipment

Archival quality

Health and safety in papermaking

All too often artists, craftspeople and students find that they do not know how the materials they work with will affect their health or what steps and precautions to take to make their studio or workshops safe places to work in, or even how to start to be environmentally conscious. Safety and health often falls awkwardly between the largely separate camps of 'art' and 'industry'. Safety in a workshop also often suffers a low profile and may appear an unattractive matter to approach. However, we urge you to take the precautions necessary to ensure your own safety and health while making paper.

Hazards in hand papermaking

The excerpts below were taken from *Art Hazard News* published by the Centre for Safety in the Arts (see SUPPLIERS). These potential hazards in the papermaking process have been gathered by Michael McCann of CSA.

- Some wood and plant materials can cause allergic reactions and skin irritation.

- The alkaline soda ash and lye are highly corrosive upon skin and eye contact, inhalation and ingestion. Boiling solutions of these alkaline materials can be very dangerous because of the risk of boiling over and the fact that steam will contain trapped alkali.

- Chlorine bleach is a skin, eye and respiratory irritant.

- Beaters can be severe safety hazards due to the chance of trapping hands in the blades when cleaning pulp from them. In addition, beaters can present noise hazards.

- The presence of large amounts of water presents electrical hazards if it splashes onto electrical outlets or other electrical equipment. In additional there is the possibility of major water leaks.

- Some pigments can be hazardous. Titanium dioxide, which is commonly used, has no significant hazards. (For hazards of other pigments, please refer to CSA's *Art Painting and Drawing* sheet).

CONSULT Your local Health & Safety Organisation (there will be a body in each country) where advice can be given about specific workshop hazards as well as chemical hazards.

NOTE Material Safety Data sheets contain information on all chemicals and pigments sold and should be always available, often upon request, wherever you buy supplies.

NOTE We have given examples of what some of the major papermaking suppliers list in their catalogues under safety and health. Alternatively, most requisites such as gloves, aprons, goggles, etc. can be obtained from specialist Safety Equipment suppliers. For your nearest supplier, consult your local telephone book.

Safety equipment

CANADA

■ THE PAPERTRAIL
1546 Chatelain Avenue
Ottawa, Ont. K1Z 8B5
Tel: (613) 728 4669
Fax: (613) 728 7996
Orders: 1 800 363 9735

Aprons
Lightweight nylon aprons with sewn edges; small and large sizes.

Barrier Cream
Provides up to 4 hours protection without gloves.

USA

■ ART HAZARDS NEWS
CENTRE FOR SAFETY IN THE ARTS
5 Beekman Street, Suite 1030
New York, NY 10038
Tel: (212) 227 62290

Contact Michael McCann, editor of *Art Hazards News*. This centre also publishes health and safety sheets on all aspects of art safety. Michael McCann has written several book about health and safety in the arts. Recommended is *Artists Beware* published by Lyons & Burford, Publishers. (See BOOKS)

■ MAGNOLIA EDITIONS
2527 Magnolia Street
Tel: (510) 839 5268
Fax: (510) 893 8334

Gloves
Chemical/oil-resistant PVC (polyvinyl chloride) coated acid gloves which come with a rough finish to give you excellent wet grip. These gauntlet cut gloves (14in) are fully coated with fleeced interlock liners and extended cuffs for roll backs.

■ LEE S. McDONALD, INC.
P.O. Box 264
Charlestown, MA 02129
Tel: (617) 242 2505
Fax: (617) 242 8825

Aprons
A lightweight and waterproof vinyl, 48 x 35in wide; wrap-around ties and neck straps.

Barrier Cream
Used to protect your hands from pigments or chemicals in wet or dry environments.

Dust Masks
Dust masks provide protection from harmful dust particles present when using chemicals, pigments and other papermaking supplies. These dust masks are NIOSH/OSHA approved and have a formable nosepiece and inner foam seal with two elastic bands for a proper fit.

Gloves
Made from natural rubber latex, these gloves are resistant to most acids and chemicals, yet provide tactile sensitivity. Yellow, 13in long, in small, medium and large sizes.

■ PRO CHEMICAL AND DYE, INC.
PO Box 14
Somerset, MA 02726
Tel: (508) 676 3838
Fax: (508) 67 -3980

Dust Respirator
NIOSH/OSHA approved respirator, made of non-woven fibres with contour fit.

Gloves
Highly-flexible, lightweight, industrial-quality, 12in long gloves which will resist puncture and abrasion. Flock lined for comfort.

Safety Goggles
Deep, well-proportioned frame fits easily and comfortably over prescription glasses. Model 1012 designed for maximum protection against chemical splash.

■ TWINROCKER HANDMADE PAPER
PO Box 413
Brookston, IN 48923
Tel: (317) 563 3119
Fax: (317) 563 8946

Aprons
These are white, lightweight, industrial-quality aprons to keep you dry at the vat, and help protect you and your clothing from pulp, pigments, sizing, etc.

Ear Plugs
These individual earplugs provide comfortable hearing protection when working near a Hollander beater or other noisy equipment. The soft foam plugs are connected together with a lightweight cord so they hang around your neck when not in use (to prevent you from losing just one).

Kerodex Barrier Creams
Barrier creams provide skin protection without interfering with your sense of touch. Use when working with pigments and other additives.
1 '71' helps protect your skin from water-based materials.
2 '51' helps protect your skin from oil-based materials.

Safety Glasses
These protect your eyes from splashing when diluting sizing and other solutions.

Papermaking and the environment

Paper drying in Pondicherry, India

In his seventeen years of supplying materials for hand papermaking, Lee S. McDonald writes, he has witnessed an evolution "beginning with the revival of hand papermaking (almost a lost art) to the present when papermaking concerns have expanded to embrace the ecosystem".

Our society is going through a revolution. Never before have ordinary people been so aware of the need to preserve our natural resources and enviroment. In the past few years we have seen the recycling of paper develop, presenting an example of how market demands have led to a wider availability. The onus of awareness and action is on papermakers themselves to look for and use less environmentally damaging processes and substances. This inevitably is going to mean experimenting with new materials and different processes.

Paper testing equipment

It is possible by simply looking and touching a sheet of paper to determine much about its manufacture. However, several important features will not be revealed to the naked eye - purity or acid content, durability, lightfastness, dimensional stability, tensile strength, etc. These factors often relate to the length of life of a paper as well as its ability to withstand the various stresses and strains of use. To verify a manufacturer's statements, certain methods of paper testing are available which will indicate or detect harmful elements such as alum or groundwood pulp or determine how a fibre has been beaten, etc. If you wish to have paper tested or indeed to buy the testing products themselves then it is best to seek professional help.

NOTE Each country will have a defined set of standards and a number of testing authorities. Seek them out via your paper or archival supplier.

NOTE Paper mills and fibre suppliers may also have testing equipment.

Most of the paper made while learning will have to be repulped.

J Barcham Green *Papermaking by Hand* in 1953 (Out of print)

CANADA

■ CRANBERRY MILLS
RR No. 1
Seeleys Bay
Ontario K0H 2N0
Tel: (613) 387 1021

Pulp and paper testing instruments available include Canadian Standard Freeness Tester, Basis Weight, Motorized Caliper Micrometer, Bending Length (Stiffness) Tester, pHep Meter and Accutint pH Indicator Papers, etc. Contact Ted Snider for more details.

SWEDEN

■ LORENTZE & WETTRE
Box 4
S-16493 Kista

Testing instruments. Write for details.

UK

■ MESSMER INSTRUMENTS LTD.
Unit 1, Imperial Business Est.
West Mill, Gravesend
Kent DA11 0DL
Tel: (01474) 566488
Fax: (01474) 560310

Messmer Instruments (with sister company Buchel van der Korput, Veenendaal, Netherlands) are specialist designers and manufacturers of physical testing equipment for the pulp and paper industry. They also distribute complementary products from other manufacturers.

Archival quality

Archival quality is a term that suggests that a material or product is permanent, durable or chemically stable and can therefore resist chemical deterioration, and has been made to a certain standard. Papermakers may need access to an archival supplier for a range of information and products. For specific conservation or preservation questions we suggest you refer either to your papermaking supplier or to a specialist archival supplier. Papermaking organisations such as those listed in INFORMATION RESOURCES can also help with queries on archival matters. Societies and centres for conservation exist in every country - such as the American Library Association (Technical Services) or Centre for Conservation and Technical Studies at Harvard. Organisations such as these can give advice and help with archival queries.

Archival terminology

ACID
In chemistry, a substance capable of forming hydrogen ions when dissolved in water. Acids can weaken cellulose in paper, board and cloth leading to embrittlement.

ACID FREE
In chemistry, materials that have a pH of 7 or higher. Sometimes this term is incorrectly used as a synonym for 'alkaline' or 'buffered'. Acid free materials can be produced from virtually any cellulose fibre source (cotton and wood among others) if measures taken during the manufacture to eliminate active acid from the pulp.

ACID MIGRATION
The transfer of acid from an acidic material to a less acidic or pH neutral material.

ALKALINE
Alkaline substances have a pH over 7. They may be added to materials to neutralise acids or as an alkaline buffer for the purposes counteracting acid that may form in the future. While a number of chemicals may be used to form buffers, most common are magnesium carbonate and calcium carbonate.

ALPHA CELLULOSE
A form of cellulose derived from cotton. The presence of alpha cellulose in paper or board is one indication of its stability and longevity.

ARCHIVAL
A term that suggests that a material or product is permanent, durable or chemically stable and can therefore be used for preservation purposes.

CONSERVATION
The treatment of works of art or materials to stabilise them chemically or strengthen them physically, sustaining their survival as long as possible.

DEACIDIFICATION
A common term for a chemical treatment that neutralises acid in a material such as paper.

LIGNIN
A component of the cell walls of plants that occurs naturally along with cellulose. Lignin is largely responsible for the strength and rigidity of plants but its presence in paper and board is believed to contribute to chemical degradation.

NEUTRAL — Having a pH of 7; neither acid nor alkaline.

pH — In chemistry, pH is the measure of the concentration of hydrogen ions in a solution which is a measure of the acid or alkaline. The pH scale runs from 0 to 14. 7 is pH neutral; numbers below 7 indicate increasing acidity with 1 being most acid. Numbers above 7 indicate increasing alkaline content with 14 being most alkaline.

PERMANENCE — The ability of a material to resist chemical deterioration. Permanent paper usually refers to a durable alkaline paper that is manufactured according to a certain standard. Permanent papers even so depend on proper storage conditions.

PRESERVATION — Activities associated with maintaining library, archival or museum materials for use either in their original condition or in some other format. Preservation is a broader term than conservation.

Archival products

GERMANY

- ANTON GLASER
  Postfach 939
  Theodor Heuss Strasse 349
  D-7000 Stuttgart

  Paper retailer selling wide range of products including archival materials.

UK

- ATLANTIS EUROPEAN LTD.
  146 Brick Lane
  London E1 6RU
  Tel: (0171) 377 8855
  Fax: (0171) 377 8850

  Archival quality materials, equipment and accessories for conservation, restoration and preservation. Distributors in Belgium, France, Germany, Greece, Italy, New Zealand, Portugal, Spain and Switzerland.

- CONSERVATION RESOURCES (UK) LTD.
  Unit 1, Pony Road
  Horspath Industrial Estate
  Cowley, Oxford OX4 2RD
  Tel: (01865) 747755
  Fax: (01865) 747035

  A comprehensive selection of archival materials, equipment and storage options for conservation, restoration and preservation. A new material which actively assists in long term preservation of artifacts is currently available. Please write for catalogue.

- PRESERVATION EQUIPMENT LTD.
  Shelfanger Diss
  Norfolk IP22 2DG
  Tel: (01379) 651527
  Fax: (01379) 650582

  Comprehensive selection of archival quality materials, equipment and accessories for conservation, restoration and preservation. UK outlet for the large University Products Inc. Catalogue.

USA

■ CONSERVATION RESOURCES INTERNATIONAL, INC.
8000 H Forbes Place
Springfield, Virginia 22151
Tel: (703) 321 7730
Fax: (703) 321 0629

A comprehensive selection of archival materials, equipment and storage options for conservation, restoration and preservation. A new material which actively assists in the long term preservation of artifacts is currently available. Please write for further details.

■ TALAS
568 Broadway
New York NY 10012
Tel: (212) 219 0770
Fax: (212) 219 0735

A supply house specialising in archival and conservation materials. Also includes tools, chemicals and adhesives. A catalogue is available for mail order. Visits are encouraged as many products sold are on display. Contact Jake or Marjorie Salik.

■ UNIVERSITY PRODUCTS, INC.
517 Main Street
PO Box 101
Holyoke, MA 01041-0101
Tel: (413) 532 9431
Orders: 1 800 336 4847

Established in 1968, this is one of the largest manufacturers and distributors of safe materials for the preservation of documents and artifacts in archives. Storage boxes, specimen trays, acid free tissue papers, adhesives, etc. Archival quality materials for conservation, restoration and preservation. Write for catalogue.

As many workers know to their cost, small is not always beautiful.

Norman Willis, General Secretary, Trades Union Council (UK) 1989

The basic idea underlying paper art . . . is that by shaping and designing the pulpy substrate, one is working counter to the traditional concept of paper as a neutral support.

Left Laurence Barker in *Internationale Biennale der Papierkunst*, Leopold Hoesch Museum, Düren, Germany 1986
Above right Helmut Frerick with his pulp sprayer

Paper casting
Vacuum systems
Pulp spraying equipment

Paper casting, Vacuum systems, Pulp spraying equipment

One of the most dramatic developments in hand papermaking is the growing awareness of paper's sculptural potential. The range of visual and tactile qualities possible through making paper by hand is dramatic. Once the pulp is made, alternatives will proliferate. There is an array of techniques for patterning and colouring pulp, spraying, stencilling, pouring as well as adding embedments or layers of colour, making embossments or casting freehand plus any form of manipulation. Casting usually begins with the duplication of a (positive) form as a (negative) mould. Most cast paper is made by pressing prepared pulp, or by laminating small pieces of lightly pressed paper, into a plaster or latex mould. When the paper has dried, it is separated from the mould and functions on its own as a bas-relief. Paper can be hand-moulded or cast to assume almost any shape and size and still be lightweight. Recent developments include pouring pulp into a flexible fabric screen mould and spraying pulp over three-dimensional armatures. The equipment used for pulp spraying is based on an industrial tool used for spraying paint, plaster or concrete, powered by an air compressor. Vacuum forming techniques have also opened up a range of possibilities which include large-scale relief pieces with a wide variety of surface textures. All vacuum systems work by using atmospheric pressure to compress a layer of pulp, thereby extracting water from it. Most systems consist of a waterproof table surface, with small holes drilled at regular intervals; an air tight chamber beneath the table into which the water is drawn; a holding tank for collecting the water and a vacuum pump. To create the vacuum, a plastic sheet is laid over the wet pulp and a seal is formed between the plastic and the table surface. When the vacuum pump is turned on it removes the air from under the plastic sheet which then allows the atmospheric pressure to press down on the paper.

CANADA

■ KAKALI HANDMADE PAPERS INC.
1249 Cartwright Street
Vancouver, B.C. V6H 3R7
Tel: (604) 682 5274

Vel-Roc
Quick drying, plaster gauze bandages for facial casts and mouldmaking.

■ THE PAPERTRAIL
1546 Chatelain Avenue
Ottawa, Ont. K1Z 8B5
Tel: (613) 728 4669
Fax: (613) 728 7796
Orders: 1 800 363 9735

Pattern Pulp Sprayers
Pattern pistol and hopper for spraying pulp onto armatures. Compressor not supplied.

GERMANY

■ HELMUT FRERICK
Artishof
Berg
D-52385 Nideggen
Tel/Fax: (24 27) 1695

New European Pulp Spraying System
Spraying paper pulp is an 'exciting technique' and this system allows you to make large thin sheets of pulp onto huge sheets of fabric and to spray multiple layers to make thick three-dimensional works. Also custom-made vacuum systems and diverse tools for hand papermaking. (See illustration p 83)

Gravity Spray Guns
This not only has a pulp flow regulator, but the airflow can also be regulated very precisely. Tank can handle up to 5 litres of pulp.

Portable Compressors
The small compressor is a high-performance, twin-head, portable compressor for professional work.

NETHERLANDS

■ PETER GENTENAAR
Churchillaan 1009
2286 AD Rijswick
Tel: (1742) 6961

Vacuum Tables
A versatile, vacuum system designed and built to customer specification.

UK

■ TIRANTI
27 Warren Street
London WC1
Tel: (01734) 312775
Fax: (01734) 323487

Sculptor's materials for casting, mould making, modelling and carving, etc.

. . . my adventures and inventions began.

Golda Lewis in *Internationale Biennale der Papierkunst*, Leopold Hoesch Museum, Düren, 1986

USA

■ CARRIAGE HOUSE PAPER
79 Guernsey Street
Brooklyn, New York 11222
Tel/Fax (718) 599 PULP (7857)
Orders: 1 800 669 8781

Pattern Pistol Sprayers
Pulp Spraying Kit includes an aluminium pistol gun sprayer supplied with a 1.75gal plastic hopper which will not rust or break, two interchangeable and adjustable nozzle fittings for a variety of spraying patterns plus a special release valve to turn the pressure off/on.

Stainless-steel Pulp Sprayers
A new pulp sprayer made in Europe. Easy to handle and clean with stainless-steel gun element. Three different nozzles that can be exchanged without the need of tools.

Charles Hilger Vacuum System
A unique system designed to be set up on any waterproof surface. Includes a vacuum pump, pressure gauge, collecting tank for the extracted water, hoses, connections and instructions (with tape and photographs) of how to use. Ideally suited for embossed forms, large pieces, works in relief, irregular shapes, uneven surfaces, pulp painting and generally all situations where an artist would have to sponge out all excess water manually.

Polyester Fabric
A non-woven material similar to interfacing fabric. Excellent as a support material in vacuum table work and pulp spraying. Available in three weights and sold by the linear yard.

The Pulp Sprayer System (See LEN KEL MANUFACTURING)

■ GOLD'S ART WORKS, INC.
2100 North Pine Street
Lumberton, NC 28358
Tel: (919) 739 9605

White Hydrocal Plaster: A strong durable plaster for making moulds; Sculpta Mould: For casting and sculpting, models like clay and sets hard in 30 minutes. Can be sanded, sawed and nailed; Marlex: Self-hardening clay in moist form; Cemco Mould Builder: Liquid latex rubber; PVA Mould Release: A water soluble liquid release agent used on the inside of moulds to ease removal of castings.

■ LEN KEL
MANUFACTURING
PO Box 2992
San Rafael, CA 94912
Tel: (510) 232 0949

The Pulp Sprayer Systems
Specifically designed for papermakers, sprays any fibre quickly and evenly (up to 1.5gal of wet pulp per minute). Kit includes pulp gun with trigger lock, wide and narrow nozzles for spraying a variety of patterns and a 6ft hose to remote 1gal plastic hopper. Extra hoppers available. Compressor not supplied.

■ MAGNOLIA EDITIONS
2527 Magnolia Street
Oakland, CA 94607
Tel: (510) 839 5268
Fax: (510) 893 8334

The Pulp Sprayer Systems (see LEN KEL MANUFACTURING)

Hilger Universal Vacuum Systems (see CARRIAGE HOUSE)

Also Latex Rubber: A rubber compound used to make flexible moulds for paper caster. Good for multiple castings with a plaster mother mould; Plaster Bandages: Used to make moulds for paper casting (and broken limbs).

■ LEE S. McDONALD, INC.
PO Box 264
Charlestown, MA 02129
Tel: (617) 242 2505
Fax: (617) 242 8825

Pattern Pistol Sprayers
Kit includes pistol gun sprayer supplied with 1.75gal plastic hopper.

Wet-Dry Vacuum Systems
An easy-to-use system that works on the top of any waterproof surface in conjunction with a wet-dry vacuum cleaner. Surface size 36 x 40in. Hose attachment and sealing tape for vacuum hose supplied, plus instructions.

Universal Vacuum Systems
Designed for use on any flat, waterproof surface, it can be used for any deep embossment, collage, assemblage, etc. Comes with vacuum pump, collection chamber, seals, hoses, etc. plus foam, screen, mesh, PVC canopy and felt materials plus a cassette tape and photographs as instruction.

Custom Vacuum Systems
Lee S. McDonald designs and builds specialised vacuum systems for artists and production papermakers. These can be flat, table-top models or drop-box configurations, powered by wet-dry vacuum cleaners or high-capacity pumps.

Ultra Felt II: A strong, very thin, white, non-woven, 100% polyester material. Used to support wet paper on a vacuum table. Not recommended as a couching felt.

TO MAKE A PAPERMAKER'S HAT

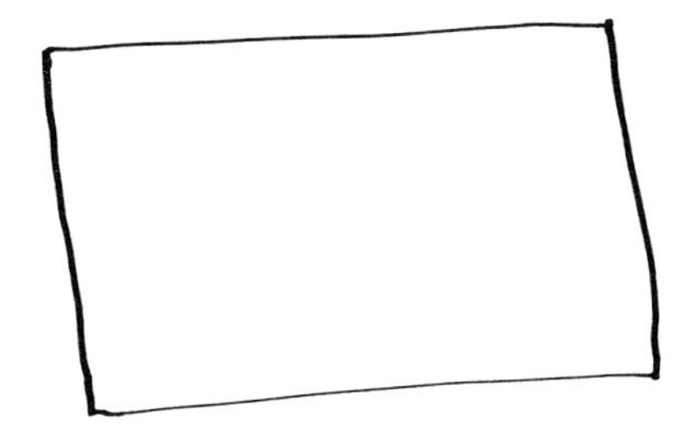

1 Take one page of newspaper (or any paper)

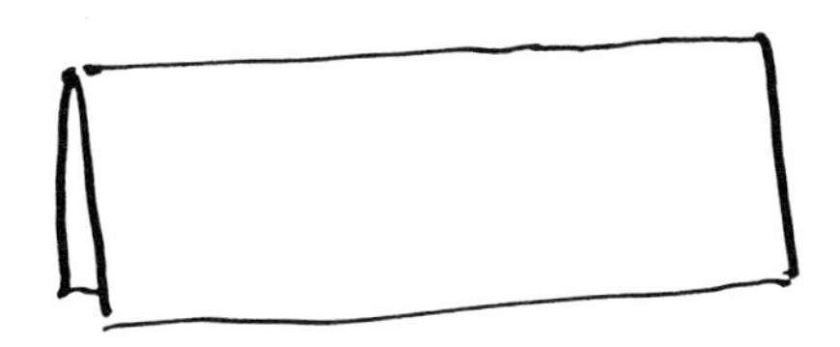

2 Fold it in half lengthways

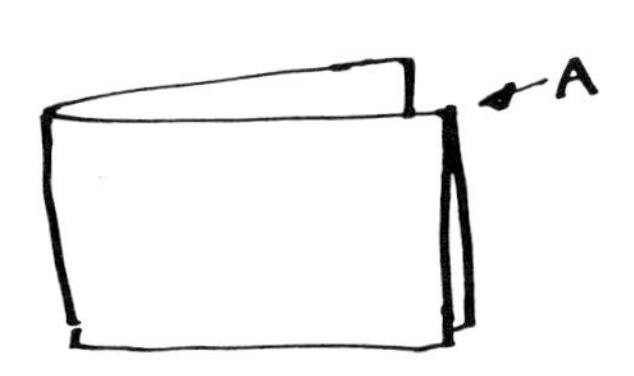

3 Fold it in half again

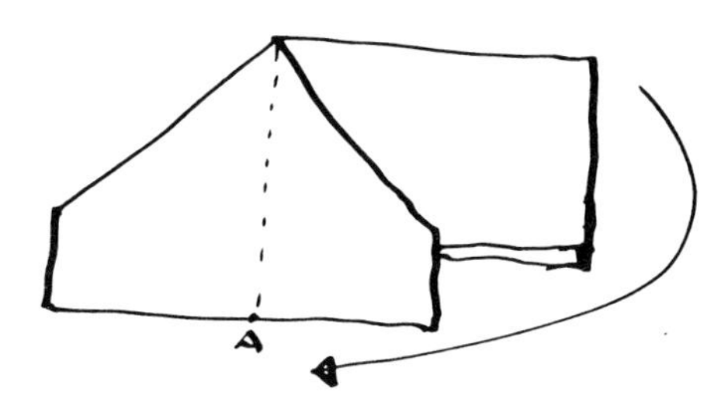

4 Move point A down by opening out the top edge folds and make a new fold

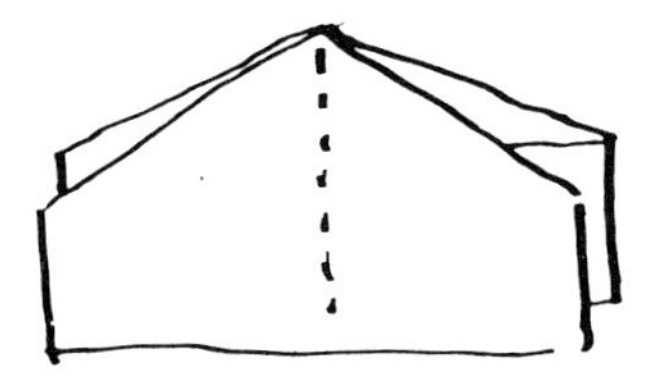

5 Turn the piece over and form a new corresponding triangle

6 Fold bottom edges once

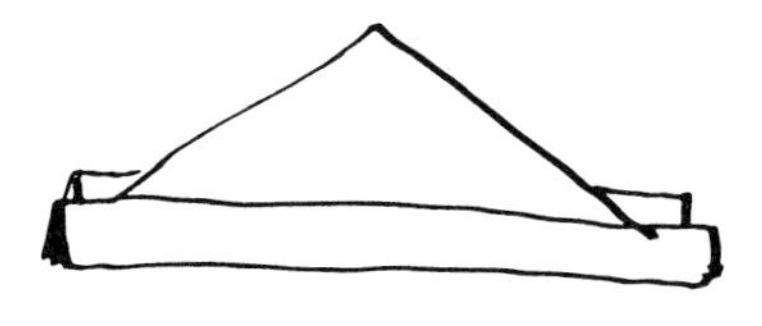

7 Then again

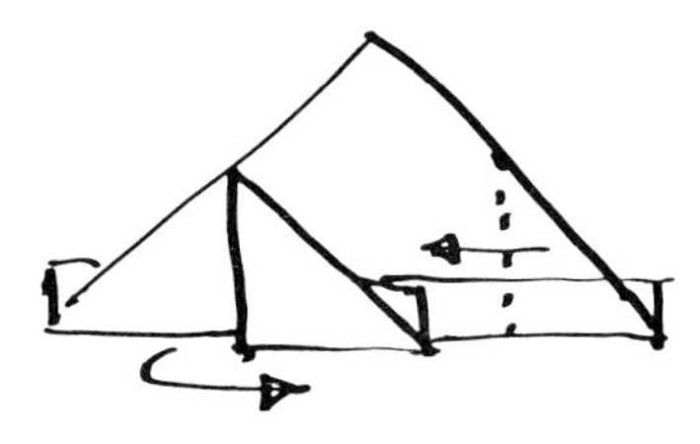

8 Fold bottom corners forward and then tuck rectangular ends inside each other

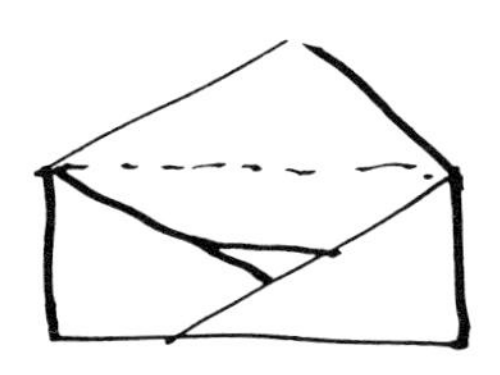

9 Make new creases by folding down the apex

Papermaking kits

Videos

Magazines

Books

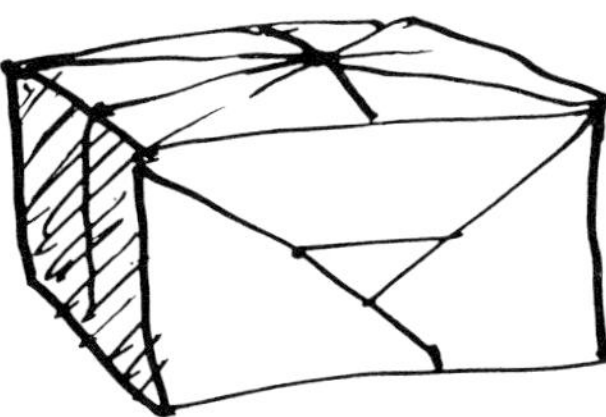

10 Put hands inside and push outwards to make finished hat (See p90).

Papermaking kits

ARGENTINA

- EL MOLINO DEL MANZANO
  Fondo de la Legua 375
  1642 San Isidro
  Buenos Aires
  Tel: (1) 763 8682

Papermaking Kits
Includes a mould and deckle, 20 felts (synthetic), 0.5kg linters plus a 'butterfly' press.

AUSTRALIA

- PAPER CAPERS
  De-Arne King
  PO Box 281
  Neutral Bay NSW 2089
  Tel: (02) 964 9471

Papermaking Kits
Two sizes available, A4 or A5, made from laquered Australian Plantation Radiata pine (recycled timber also employed) with a fibreglass mesh screen (polypropylene heat shrinking mesh also available). The papermaking kit is packaged in a recycled Carry Bag and includes a template which enables you to make three different envelopes. With Instructions which are also sold separately. Also Envelope Mould and Deckle.

- THE PAPER MERCHANT
  316 Rokeby Road
  Subiaco, WA 6008
  Tel: (09) 381-6489
  Fax: (09) 381 3193

Papermaking Kits
Full range, with instructions.

CANADA

- KAKALI HANDMADE PAPERS, Inc.
  1249 Cartwright Street
  Vancouver, B.C. V6H 3R7
  Tel: (604) 682-5274

Papermaking Kits
Contains a 4 x 5.5in mould and deckle , foam felts, cellulose sponge, sizing, handmade paper sample, fully-illustrated instruction booklet and enough cotton and abaca fibres to make at least 60 sheets.

- THE PAPERTRAIL
  1546 Chatelain Avenue
  Ottawa, Ont. K1Z 8B5
  Tel: (613) 728 4669
  Fax: (613) 728 7796
  Orders: 1 800 363 9735

Small Papermaking Kits
Contains a 5 x 7in mould and deckle, 125ml sizing, 0.5kg unbleached abaca, 0.5kg 1st cut cotton linters, 8 traditional papermaking felts (12 x 14in) or 20 interfacing-type felts, plus instructions for use.

Large Papermaking Kits
Contains an 8.5 x 11in mould and deckle, 250ml sizing, 1kg unbleached abaca, 1kg 1st cut cotton linters, 8 traditional papermaking felts (12 x 14in) or 20 interfacing type felts, plus instructions for use.

■ LA PAPETERIE SAINT-ARMAND 3700 Saint Patrick Street Montreal, Quebec H4E 1A2 Tel/Fax: (514) 931 8338	Papermaking Kits Hand mould for 6 x 9in sheets plus felts and instructions.

UK

■ PAPER PLUS Maggie Holland 24 Zetland Road Chorlton, Manchester 21 2TH Tel: (0161) 881 0672	Papermaking Kits Contains instruction sheet, postcard-sized mould and deckle, beaten pulp, pressing boards, couching cloths. Japanese Papermaking Educational Resource Packs (see ORIENTAL PAPERMAKING)
■ HOPE EDUCATION LTD. Orb Mill, Huddersfield Road Waterhead, Oldham OL4 28T Tel: (0161) 633-6611	Papermaking Pack Q5974/001 This kit provides all you need to make paper from waste paper includes posters, instructions for papermaking and teacher's notes.
■ SPECIALIST CRAFTS LTD. (Formerly DRYAD EDUCATIONAL DIVISION) PO Box 247, Leicester LE1 9QS Tel: (0116) 2510405 Fax: (0116) 2515015	Decorative Papermaking Kit RP5220 Contains mould and deckle, pulp, couching cloths, marbling colour and pipettes, paper press and instructions for use. Contact Paul Crick.

USA

■ CARRIAGE HOUSE PAPER 79 Guernsey Street Brooklyn, NY 11222 Tel/Fax: (718) 599 PULP (7857) Orders: 1 800 699 8781	Papermaking Kits Kits with (or without) a vat (plus a lid which is made from heavy-gauge plastic), a mould and deckle, 12 papermaking felts, enough cotton linters and abaca pulp to make 150 sheets of paper, a 4oz jar of sizing and an instruction booklet. Available in 5.5 x 8.5in; 8.5 x 11in.
■ GREG MARKIM, INC. PO Box 13245 Milwaukee, WI 43212 Tel: (414) 453-1480 Fax: (414) 453-1495	Original Tin Can Papermaking Kit (101) This kit contains support screen, water removal screen, 12 couch sheets, sponge, press bar and instructions. It will make up to 6 sheets in one sessionand all items are reusable. Round sheets up to 5in diameter.

Papermaker with hat

Original Plus Tin Can Papermaking Kit (102)
As in Original Tin Can (over) plus a drain rack, 12 additional couch sheets and an envelope pattern guide.

Classroom Tin Can Papermaking Kit (103)
Supplies to make up to 35 sheets per session. Includes a set-up guide, instructions and envelope pattern guide.

Clear Cylinder (111)
'See through' cylinder replaces the top can in Tin Can Papermaking. Useful for teaching demonstrations.

Tin Can Couch Sheets
From the Classic Hand Papermaking Series.

Classic Hand Papermaking Kits
Classic Hand Mould (makes 6 x 8in paper) including 1 screen, 1 water removal screen, 50 couch sheets (reusable), press bar, sponge and drain rack, etc.

Classic Classroom Papermaking Kits
2 screens, etc. 100 couch sheets. Other items same as for Classic Papermaking Kit.

Professional Hand Papermaking Kits
Quantities same as Classic Papermaking Kit, making 8.5 x 11in paper.

Professional Classroom Papermaking Kits
Professional Hand Mould (making 8.5 x 11in paper). Other items and quantities the same as Classic Classroom Papermaking Kit.

■ GOLD'S ARTWORKS, INC.
2100 North Pine Street
Lumberton, NC 28358
Tel/Fax (919) 739 9605
Orders: 1 800 356 2306

A papermaking kit for beginners, designed in 1983, with easy to follow instructions. Sheet size 5 x 7in. Kit contains a mould and deckle, cotton linter pulp, sizing, synthetic chammy.

■ LEE S. McDONALD, INC.
PO Box 264
Charlestown, MA 02129
Tel: (617) 242 2505
Fax: (617) 242 8825

Student Papermaking Kits
A 5.5 x 8.5in student mould and deckle, 4 papermaking felts size 8.5 x 11in, 1lb cotton linters, 0.25lb abaca pulp plus Instruction Booklet (which takes you through the process).

Basic Papermaking Kit,
8.5 x 11in basic mould and deckle, 4 papermaking felts size 12 x 14in, 2lb cotton linters, 0.5lb abaca pulp plus Instruction Booklet. Both kits are also available with a vat.

Kit in a Vat
The Kit in a Vat has a small black plastic vat (see VATS) plus a papermaking kit.

Package Deal for Schools
A Package Deal for Schools with all the essentials to make paper for a class of about 18 students. Includes 6 Student Moulds (5.5 x 8.5in), 4 papermaking felts, 10lbs cotton linters, 5lbs abaca pulp, 1 pint of liquid sizing, 1 vat plus *A Guide to Sheet Forming and Paper Casting*.

Package Deal with Ready-To-Use Pulp
The same Package Deal is available with 2 containers of ready-to-use pulp. Each 5gal container includes of 2lb of sized and drained cotton pulp.

■ TWINROCKER HANDMADE PAPER
PO Box 413
Brookston, IN 47923
Tel: (317) 563 3119
Fax: (317) 563 8946

Classroom Package
This package contains the book *Paper by Kids* by Arnold Grummer; 2 flat deckle moulds (one large and one small); 12 felts; 2oz jars of black, red, yellow and blue pigments; choice of 5gal pail of ready-to-use 27 or 29 pulp (medium beating, with sizing).

Hand Papermaking Kits
Ribless mould with a simple, flat deckle, 4 felts, a basic Instruction Booklet and a choice of 1lb of abaca half-stuff or 1lb of cotton linters half-stuff, or a 1gal pail of ready-to-use cotton linters pulp. Either 5.5 x 8.5in or 8.5 x 11in.

Arnold Grummer Papermaking Kits
Designed to accompany *Paper by Kids* written by Arnold Grummer. The two sizes are 6 x 8in paper or 8.5 x 11in paper. These kits include a direction sheet (pictures and text); a 'pour' hand mould; 2 professional papermaking screens; 2 water removal screens; drain rack; 100 couch sheets; sponge; press block; pattern for making envelopes; information packet on how to get pulp and taking care of your equipment.

Arnold Grummer Tin Can Papermaking Kits
Developed to accompany Arnold Grummer's latest book, *Tin Can Papermaking* and designed to recycle paper easily using two tin cans. The kit contains: a papermaking screen; support screen, water removal screen; 12 couch sheets, press block; sponge and instructions. You have to supply your own tin cans.

Videos

USA

- CARRIAGE HOUSE PAPER
79 Guernsey Street
Brooklyn, NY 11222
Tel/Fax: (718) 599 PULP (7857)

Elaine Koretsky has documented some of her extraordinary papermaking experiences in Asia. Two videos are available, each recording a highly specialised method of papermaking - *The Last Papermakers on the Silk Road,* filmed in Khotan, Xinjiang Province, China and *Papermaking in Sechuan Province.*

- HAND PAPERMAKING
SLIDE KITS & VIDEO
PO Box 77027
Washington, DC 20013-7027

Artists' work represented in a slide carousel accompanied by a cassette of artists' statements. This is also available on video format. Please apply to Hand Papermaking for further information on rental or orders (see below).

- GREG MARKIM INC.
PO Box 13245
Milwaukee, WI 53213
Tel: (414) 453 1480
Fax: (414)453 1495

Papermaking for Everyone Video. This video explains how to make paper by hand, how to recycle, how to make paper with two tin cans, how to dry paper with and without heat and many helpful tips for successful hand papermaking. VHS 57 minutes long.

- McGOWAN FILM & VIDEO
4926 Wolcott Ave
Chicago, IL 60640
Tel: (312) 271 0793

The Mark of the Maker. A film, available in VHS video or 16mm film, that concerns hand papermaking at Twinrocker, written, produced, directed, filmed and financed by David McGowan and Laurie Kennard, "much more than a papermaking how-to-do-it; it is also a why-we-do-it" at the Twinrocker mill.

- PAPERMAKING VIDEO TAPES
University of Iowa
Center for the Book
364 English Philosophy Building
Iowa city, Iowa 52242
Tel: (319) 335 0438
Fax: (319) 335 2535

Written by and featuring Timothy Barrett, these instructional videos emphasise the mechanics of papermaking and show the papermaking processes step-by-step. They are also available in foreign language versions. Please contact the Translation Laboratory (in any language). They include *Japanese Style Papermaking I, II, and III* - a very comprehensive set varying from 36 to 80 minutes in length. Also *Western Papermaking,* which explains classroom equipment and techniques, fibre selection, beating, sheet formation, drying methods, sizing, etc. and *Western Papermaking* II, detailing professional equipment and techniques. Both approximately 25-30 minutes in length.

- THE PULPERS
1101 North High Cross Road
Urbana, IL 61801
Tel: (212) 328 0118

A division of Editions in Cast Paper Ltd. Video series *Hand Papermaking I-IV* (VHS, approximate running time - 10 minutes each).

Magazines

It is often essential to keep in contact with news, views and developments of others in the papermaking field. We have listed the specialist publications that we know of but would be pleased to hear of any others for future editions. Many papermills and workshops produce their own newsletters (e.g. Dieu Donné mill in New York). Write directly to the mills for information.

NOT LISTED We have not listed the paper bulletins, magazines and newsletters of societies such as IAMPA, IPH, FDH, etc. These are important sources of information on many levels (see INFORMATION RESOURCES).

NOTE Papermaking articles and coverage is also often found in specialist art magazines and newsletters of crafts, fiber arts, printmaking, printing, paper trades, bulletins of paper conservationists, etc. Again, these are not listed here.

- HAND PAPERMAKING
  PO Box 77027
  Washington, DC 20013-7127
  USA

Co-founded by Michael Durgin, *Hand Papermaking* is a semi-annual magazine with a quarterly newsletter and a must for anyone involved with hand papermaking. *Hand Papermaking* is a non-profit organisation devoted to advancing traditional and contemporary ideas in the art of hand papermaking through publications and other means. The magazine discusses innovative new techniques in paper medium, documents current papermakers, mills and processes, paper decorators, etc. including essays and samples from many and varied sources. Subscription on an annual basis. Write for further information.

- PAPER CRAFTERS
  6575 SW 86th Street
  Portland OR 97223
  Phone/Fax: (503) 223 0167

A quarterly journal for paper artists. Write for further details.

- FIBERARTS
  50 College Street
  Asheville, NC 28801
  Tel: (704) 253 0467

Fiberarts is an American magazine which provides and overview of the fiber art field. Full colour with special issues focusing on different aspects of fiber art including tapestry, papermaking, rugs, etc.

- AMERICAN CRAFT
  American Crafts Council
  40 W. 53rd street
  New York, NY 10019
  Tel: (212) 956 3535

This is a high quality, full-colour, American magazine published bi-monthly by the American Crafts Council and serves as a showcase for contemporary professional crafts including papermaking. Write for more information.

- CRAFTS
  Crafts Council
  44a Pentonville Road
  London N1 9BY
  Tel: (0171) 278 6891
  Fax: (0171) 837 6891

Produced by the Crafts Council in Britain, a bi-monthly, high quality, full-colour magazine concerning all aspects of crafts, with very occasional articles on hand papermaking. Write for further information.

Books

Barrett, Timothy *Japanese Papermaking, Traditions, Tools and Techniques* Weatherhill New York/Tokyo 1983

Bolton, Claire *AwaGami - Japanese Handmade Papers from the Fuji Mills, Tokushima* Alembic Press Abingdon 1992

Bell, Lillian A. *Plant Fibers for Papermaking* and *Papyrus, Tapa, Amate and Rice Paper* Liliaceae Press McMinnville, Oregon 1982

Cunning, Sherill *Handmade Paper: A Practical Guide to Oriental and Western Techniques* Ravens Word Press Escindido, California 1982

Dawson, Sophie *The Art and Craft of Papermaking* Aurum London 1993

Doizy, Marie Ange and Pascal Fulacher *Papiers et Moulins* Editions Technorama Paris 1989

Grummer, Arnold *Tin Can Papermaking*

Heller, Jules *Papermaking* Watson Guptill New York 1978

Hughes, Sukey *Washi - The World of Japanese Paper* Kodansha International Tokyo/New York/San Francisco 1978

Hunter, Dard *Papermaking History & Technique of an Ancient Craft* Dover New York 1978

Hunter, Dard *Papermaking in the Classroom* Oak Knoll Books Newcastle Delaware 1994 (1932)

Koretsky, Elaine *Colour for the Hand Papermaker* Carriage House Boston 1990

Koretsky, Elaine *A Gathering of Papermakers* Carriage House Boston 1988

Koretsky, Elaine and Donna *The Gold Beaters of Mandalay: Hand Papermaking in Burma* Carriage House

Long, Paulette *Paper, Art & Technology* World Print Council, San Fransisco 1979

International Biennale der Papierkunst *PaperART* Leopold Hoesch Museum Düren 1994

McCann, Michael *Artists Beware* Lyons & Burford New York 1992

McCann, Michael *Health Hazards Manual for Artists* Lyons & Burford New York 1994

Macfarlane, Nigel *A Paper Journey, Travels among the Village Papermakers of India and Nepal* Oak Knoll Books Newcastle Delaware 1993

McDonald, Lee S. *The Beater Builders of North America* 1946-89 Friends of the Dard Hunter Paper Museum Inc. Buffalo, NY 1990

Richardson, Maureen *Plant Papers* and *An Alternative to the Beater* Plant Papers Herefordshire 1990

Rudin, Bo *Making Paper - A Look into the History of an Ancient Craft* Lyons & Burford New York 1990

Saddington, Marianne *Making Your Own Paper - An Introduction to Creative Paper-making* New Holland (Publishers) Ltd. 1991

Shannon, Faith *The Art and Craft of Paper* Mitchell Beazley London 1991

Shimura, Asao *Kami '89* Cannabis Press 1992

Studley, Vance *The Art and Craft of Handmade Paper* Studio Vista; Dover New York 1978

Thompson, Claudia G. *Recycled Papers - The Essential Guide* MIT Press Cambridge Massachusetts 1992

Toale, Bernard *The Art of Papermaking* Davis Publications 1989

Turner, Silvie *Which Paper? A Guide to Fine Papers for Artists, Craftspeople and Designers* estamp London; Design Books New York 1990

Visbøll, Anne *Papirmageri - handlavet papir* (1985), *Papirmageri 2 kunst og teknik* Borgen 1990

Williams, Nancy *Paperwork - the potential of paper in Graphic Design* Phaidon London 1993

I understood something of John's joy with his work and therefore also a piece of his character after he explained about his enjoyment of playing with water . . . And then I saw him, with both arms deep in the vat, stirring and mixing up the pulp in water with the strength of his entire body. There was an energy that I had not expected of him. By now it was finally clear. This was his element.

Above From *Papierschöpfungen. Arbeitne aus der Papierwerkstatt John Gerard*
Die Deutsche Bibliotek 1994
Right John Gerard in his paper workshop

Paper workshops

Paper workshops

There are many ways to learn the techniques of papermaking. CLASSES, WORKSHOPS, SEMINARS exist in many forms in many locations in varying levels of expertise and experience. Courses and workshops vary and are individual to the teachers and the market requirements. Check the courses out well in advance of wanting to register and send a SAE for information.

. . . experience is the only genuine knowledge . . .

Goethe

One of the best ways to learn the craft of papermaking is to be TAUGHT BY A MASTER. Many well-known and experienced papermakers across the world share their knowledge and expertise with those wishing to learn this craft. This list identifies a range of different types of teaching experience but by its nature is not comprehensive nor local.

Many APPRENTICESHIPS and INTERNSHIPS, RESIDENCIES and SPONSORED PLACES exist in papermaking workshops across the world. Write for information to research the availability of these places.

Many workshops also COLLABORATE with artists to produce art works in paper. Collaborations play an important part influencing an artist's approach to papermaking. Again this is not a definitive listing (it does not include printmaking workshops with paper facilities for instance) but simply a starting place for research.

HIRE OF STUDIO or EQUIPMENT or TECHNICAL HELP also provides opportunities for papermaking.

PAPER TRAVELLERS will find that many papermaking tours and trips are organised around the world to visit known (and unknown) papermaking facilities/villages/areas. Elaine Koretsky and Asao Shimura are especially well-known for this. Note also that many paper groups such as IAPMA plan and organise paper conferences and trips. Begin searching and planning well before you wish to travel.

Many paper mills and museums will offer DEMONSTRATIONS of technique and VISITS to papermaking facilities.

NOT LISTED HERE Many intrepid papermakers and artists also run smaller (and shorter) papermaking courses and workshops in INSTITUTIONS, UNIVERSITIES, COLLEGES, ART CENTRES, PRIVATE STUDIOS, SCHOOLS, ADULT EDUCATION CENTRES, PRINTMAKING STUDIOS, MUSEUMS and many other locations everywhere. To find them look in your local and/or national art press, special papermaking bulletins, art centres, museums or ask your friends and papermaking colleagues for details.

AUSTRALIA

PRIMROSE PAPERWORKS
PO Box 152
Cremorne, NSW 2090
Tel: (02) 909 1277
Fax: (02) 953 1834
Contact Juliette Rubensohn
Primrose Paperworks was founded in 1989 by Geraldine Berkmeier, Ruth Faerber, Sherry Cook and Juliette Rubensohn. It has a comprehensive workshop schedule in papermaking and related arts and product sales.

PAPER CAPERS
PO Box 281
Neutral Bay, NSW 2089
Tel/Fax: (02) 964 9471
Contact De-Arne King
A busy company supplying entirely Australian-made paper products, pulp, presses, etc. as well as teaching the craft of papermaking. Write for details.

CANADA

ATELIER PAPYRUS INC.
Trois Rivières
Quebec G9A 1H7
Tel: (819) 372-0814
The Atelier Papyrus was established in 1984 and it provides papermaking services and facilities. The Atelier Papyrus runs workshops, events, conferences and exhibitions and other activities for papermaking.

KAKALI HANDMADE PAPERS INC.
1249 Cartwright Street
Granville Isalnd
Vancouver BC V6H 3R7
Tel:(604) 682 5274
Workshops in papermaking and book arts. Write for further information.

NEW LEAF PAPERMILL
105 Twinflower Way
Saltspring Island
BC V8K 1R4
Tel: (604) 537 5335
Workshops in papermaking in several areas. Write for further information.

THE PAPERTRAIL
1546 Chatelain Ave
Ottawa, Ont. K1Z 8B5
Tel: (613) 728 4669
Fax: (613) 728 7796
Located in an old warehouse, the mill offers workshops in all aspects of papermaking and book arts, group programmes, independent papermaking (use of facilities without technical assistance) and artist's collaboration. A third of the space serves as a special projects room and paper art gallery.

DENMARK

ANNE VILSBØLL
Fredensgade 4
Strynø
DK-5900 Rudkøbing
Tel: (62) 51 50 02
Fax: (62) 51 50 12
This was the first hand papermaking studio in Denmark, established in 1983. "The studio is involved with research and experimentation in paper as an artistic language leading to innovative uses of handmade paper as an integral part of art forms." Group visits, workshops and studio rental with assistance.

EGYPT

ASSOCIATION FOR THE PROTECTION OF THE ENVIRONMENT
Zabbeleen Paper Centre
Marie Assaad/ Isis Bailey
Manshiet Nasser, Mokattam
PO Box 32 Qalaa, Cairo
Tel: (202) 354 3305
Fax: (202) 273 5139

FINLAND

SUOMEN PAPERITAITEEN KESKUS
Finnish Paper Art Centre Association
Kokonniementie 6
FI-06100 Porvoo
Contact Väiski Putkonen, Chairman.
This organisation runs paper workshops and educational programmes in Finnish art institutes on a regular basis. Also international workshops and many other activities. Write for information.

GERMANY

HELMUT FRERICK
Pflanze-Faser-Papier
ArtisthofBerg
D-52385 Nideggen
Tel: (24 27) 16 95
Fax: (24 27) 16 95
Founded in 1986 by Helmut Frerick, paper artist, as an experimental plant paper mill. Extensive equipment, varied programme of workshops on papermaking as an art form. Also works with artists/collaborations. Own guest house available facilitating artists working and living in the same location.

PAPIERWERKSTATT
JOHN GERARD
Auf dem Essig 3
D-5308 Rheinbach-Hilberath
Tel: (22 26) 21 02
The paper workshop is one of the best and most fully equipped artists' paper studios in Europe. All necessary tools and equipment are present including over 50 moulds, hydraulic press and a Hollander beater. Started in 1985, the workshop is based on three principles: collaborative work with artists, a small sheet production facility and experimentation work with paper. John Gerard also travels widely within Germany to teach papermaking in many academies and universities.

PAPIERWERKSTATT
ALFRED KÖNIG
Friedenspromenade 31
D-81827 München

HAWAII

MARILYN WOLD
c/o PO Box 5202
Aloha, OR 97006
USA
Tel: (503) 641 2294
Fax: (503) 641 7694

Annual artists' retreat in Hawaii (and also in Oregon) with workshops in papermaking with plant fibres. Contact Marilyn Wold with SAE for details.

INDIA

SRI AUROBINDO ASHRAM
Pondicherry 605-002
Kartikeya Annurakta has been manager of the papermaking facility at the Sri Aurobindo Ashram for over 30 years. Artist collaborations. Write for details.

KHADI PAPERS INDIA
Chilgrove, Chichester PO18 9HU
Tel: (01243) 535314
Fax: (01243) 535354
Nigel Macfarlane has set up a small-scale handmade paper production 22km from Pune, Maharashtra. This is a hand papermill which concentrates on quality handmade papers and also includes a facility for artists and papermakers to visit, make paper and explore initiatives. Write for details.

ISRAEL

UNCLE BOB LESLIE PAPER MILL
PO Box 164, Omer
Tel: (011 972) 74695598
Courses in papermaking. Write for further information.

JAPAN

FUJI PAPER MILLS CO-OPERATIVE
CPO Box 114
Tokushima 770
Tel: (0886) 52 2772
Annual Japanese Summer papermaking workshop with Yoichi Fujimori.

NEW ZEALAND

VICTORIA PAPER HOUSE
Kate Coolahan
57 Sefton street
Wadestown, Wellington
Tel/Fax: (04) 4735627
Papermaking using raw and recycled fibres, casting and dyeing. Hire of studio by arrangement.

PHILIPPINES

KAMI PAPER MAKERS PHILIPPINES (KAMI PHILIPPINES)
Tina, Makato
Aklan 5611
Contact Asao Shimura. Kami Philippines started in 1993 and are specialists in making pina paper from native pineapple, *ananas comosus*. Asao also organises and runs paper workshops and tours (with flight and accommodation). Please write to him for further details.

SOUTH AFRICA

TECHNIKON NATAL
953 Durban 4000
Fax (031)2234405
John Roome, Dept. of Fine Art. Here, in the university, they run one-day workshops for the public to provide instruction in papermaking.

SPAIN

MUSEU MOLI PAPERER DE CAPELLADES
E-08786 Capellades (Barcelona)
Tel/Fax: (3) 801 28 50
Papermaking workshops. Write for further details.

SWEDEN

ZEN ART PAPER AB
St Jörgens väg 20D
S-422 49 Gothenburg
Contact Christina Bolling. Small papermill and workshops in papermaking.

SWITZERLAND

VIVAINE FONTAINE
Pres de la Fontaine
CH-1637 Charmey
Tel: (029)718 55
Workshops given throughout the year in papermaking, Japanese papermaking and sculptural techniques. Write for further details.

PAPIERATELIER
Teufenerstr.75
CH-9000 St Gallen
Tel: (071) 23 50 66
Contact Suzanne Zehnder or H. Böckle. Papermaking workshop within an adult education centre for arts and crafts. Papermaking space and equipment can be rented. German, Italian, French and English speaking.

THERESE WEBER
Atelierhaus
Fabrikmattenweg 1
CH-4144 Arlesheim
Tel/Fax: (061) 70191 07
Independent papermaking studio set up in 1988 by Therese Weber. Her interest is mainly in research, experimentation and the use of handmade paper as an art form. Artist collaborations are welcome. Studio is also available for rental.

UK

THE BRITISH PAPER COMPANY
Frogmore Mill
Hemel Hempstead, Herts HP3 9RY
Tel: (01442) 231234
Fax: (01442) 252963
A new hand papermaking facility is being set up at The British Paper Company who are based at the historic Frogmore Mill. At the time of going to press, this facility is in its infancy but BPC welcome inquiries for those wishing to learn hand papermaking.

SOPHIE DAWSON
Rainthorpe Hall
Tasburgh
Norwich NR15 1RQ
Tel: (01508) 470 618
Fax : (01508) 470 793
Two day Workshops, Intermediate and Advanced Workshops. Each workshop aims to further the craft of hand papermaking across the broadest spectrum of

art and design and to encourage quality and craftsmanship. SAE for list of programmes.

GLASGOW SCHOOL OF ART
167 Renfrew Street
Glasgow G3 6RQ
Scotland
Tel (0141)353 4500
Papermaking Summer schools offered by Jacki Parry. Contact her in the Printmaking Dept.

GILLIAN SPIRES HANDMADE PAPER
4 Fore Street
Ugborough, Ivybridge
Devon PL21 ONP
Tel: (01752) 892783
Basic course in papermaking covers traditional methods similar to those used by ancient Chinese using local plants. Also introduction to dyeing paper fibres with wood dyes and design making techniques. Write for more details.

NAUTILUS PRESS & PAPER MILL
107 Southern Row
London W10 5AL
Tel: (0181) 968 7302
Fax: (0181) 968 6309
Papermaking studio in London run by Jane Reese. Part of a complex of printing, binding and papermaking opportunities in this active studio workshop. Tuition. Courses. Send a SAE for information.

THE PAPER WORKSHOP
Gallowgate Studios
15 East Campbell Street
Glasgow 15DT
Scotland
Jacki Parry's papermaking workshop. Artist's papermaking studio for hire and collaborative work. Courses. Tuition. Enquiries welcome. Note - no sheet production or retail counter.

PLANT PAPERS
Romilley, Brilley
Herefordshire HR3 6HE
Tel: (01497) 831 546
Fax: (01497) 831 327
Maureen Richardson's Plant Papers workshop which specialises in papers from plant sources and recycling. Courses. Tuition. Workshops all year round. Local accommodation lists. Send SAE for details.

USA

ARIZONA WORKSHOPS
PO Box 1711,
Bisbee, AZ 85603
Tel: (602) 432 5924
Workshops in papermaking, bookbinding and marbling. Write for details.

TIMOTHY BARRETT
Timothy Barrett is associate resident scientist at the University of Iowa (UI) in Iowa City. As Director of UI papermaking facilities, he also teaches in the school or Art and Art history and overseas research and production papermaking at a separate Centre for the Book facility. (See VIDEOS.)

CARRIAGE HOUSE PAPER
79 Guernsey Street
Brooklyn, New York 11222
Tel/Fax: (718)599 PULP (7857)
This is a new location for Carriage House Paper and is the major teaching and papermaking facility; staff include Elaine Koretsky, Donna Koretsky, Kimberly Carr and Tin Tin Nyo. Offers rental of papermaking facilities which includes pulp spraying, plus short specialised workshops all through the year. Write for a programme.

CENTRE FOR BOOK ARTS
626 Broadway,
New York NY 10012
Tel: (212) 460 9768
Workshops in papermaking, printing and book arts. Write for details.

COLOPHON BOOK ARTS SUPPLY
3046 Hogum Bay Road NE
Olympia, WA 98506
Tel (206) 459 2940
Run by Don Guyot. Teacher of paper marbling and suminigashi. Variety of workshops, supplies and equipment for traditional and Oriental papermaking, marbling, and hand bookbinding.

COLUMBIA COLLEGE CHICAGO
CENTRE FOR BOOK & PAPER ARTS
218 S. Wabash
Chicago, IL 60604
Tel: (312) 431 8612
Classes in papermaking and book arts. Write for further information.

CREATIVE ARTS WORKSHOP
80 Audubon Street
New Haven, CT 06510
Tel: (203) 562 4927
Workshops in papermaking and book arts. Write for further information.

DIEU DONNÉ PAPERMILL, INC.
433 Broome Street
New York, NY 10013
Tel: (212) 2260573
Established in 1976, Dieu Donné Papermill (a not-for-profit arts organisation) runs workshops, classes and lectures in papermaking and book arts; offers both an apprenticeship and workspace progamme. Write for details and membership information.

DOBBIN MILL
50-52 Dobbin Street
Brooklyn, New York 11222
Tel: (718) 388-9631
Fax: (718) 388 9612
Contact Robin Ami Silverberg, Director. Situated in New York City, it comprises a large papermaking studio, an artist book studio, a darkroom and private courtyard for working outdoors. The mill makes production and custom handmade papers; offers internship and collaboration facilities; workshops and classes in hand papermaking and book arts.

FABRILE STUDIO
PO Box 1551
Taos, NM 875771
Tel: (505) 751 0306
Workshops in papermaking. Write for further information.

FARRIN & FLETCHER DESIGN STUDIOS
146 West Bellvue Dr.
Pasadena, CA 91105
Tel: (818) 796 1837
Workshops in papermaking. Write for further information.

HISTORIC RITTENHOUSETOWN
206 Lincoln Drive
Philadelphia, PA 19144
Tel: (215) 843 2228
Workshops in papermaking and book arts. Write for further information.

ICOSA STUDIO AND MILL
Route 4, Box 279
Ellensburg, WA 98926
Tel/Fax: (509) 964 2341
Contact Margaret Ahrens Sahlstrand. Hand papermill with educational programmes on Korean and Japanese papermaking and Japanese paper clothing (*shiu* and *kamiko*); other workshops.

MAGNOLIA EDITIONS
2527 Magnolia Street
Oakland, CA 94607
Tel: (510) 839-5268
Fax: (510) 893-8334
Magnolia Editions comprises a paper workshop and fine art print studio. It makes custom papers and offers collaboration facilities to artists. It is also open to the public and organises tours for institutions, schools, etc.

MINNESOTA CENTER FOR BOOK ARTS
24 N. Third Street,
Minneapolis, NM 55401
Tel: (612) 338 3634
Workshops in papermaking and book arts. Write for further information.

PAPER ARTS
Mill and studio
Tel: (602) 966 1998
'Arizona getaway with open studio'. Mill, equipment and beater rental. Call for further information.

PYRAMID ATLANTIC
6001 66th Ave
Riverdale, MD 20737
Tel: (301) 577 3424
Fax: (301) 459 7629
Pyramid Atlantic's aim is to provide an artist-centred community to foster paper, print and book arts and to promote these to the general public. Also dedicated to innovative, collaborative exchange among artists. Workshops in papermaking and book arts.

RUGG ROAD PAPERS AND PRINTS
1 Fitchburg Street, Suite B154
Somerville, MA 02143
Tel: (617) 666 0007
Workshops in papermaking. Write for further details.

BARBARA SCHECK
47 West Main Street
Treadwell, New York 13846
Tel: (607) 829 8432)
Studio visits and hands-on workshops. Write for more details.

TAOS PAPERWORKS
PO Box 3162
Taos, New Mexico 87571
Tel: (505) 758 9589
'The class will focus on sheet formation using a variety of plant fibres, how to use paper pulp as both 2D and 3D art medium and how to gather and prepare local plant fibres for use in papermaking'. Weekend workshops throughout Spring and Summer.

TOCKINGTON PAPER STUDIO
6230 N. 8th Street,
Philadelphia, PA 19126
Tel: (215) 548 9334
Workshops in papermaking. Write for further details.

TWINROCKER HANDMADE PAPER
PO Box 413
Brookston, IN 47923
Tel: (317) 563-3119
Kathryn and Howard Clark founded Twinrocker in 1971. Extensive collaborative work is done with artists and qualified apprentices are accepted for study. The mill is committed to education concerning history and process of papermaking and welcomes invitations to lecture and run workshops. Local papermaking workshops within a three hour drive from Brookston. Individual and group tours by appointment.

VISUAL STUDIES WORKSHOPS
31 Princes Street
Rochester, NY 14607
Tel: (716) 442 8676
Workshops in book arts. Write for further information.

WATERLEAF MILL & BINDERY
Pat Baldwin
PO Box 1711
Bisbee, AZ 85603
Tel: (602) 432 5924
Apprenticeships in papermaking, marbling, etc. Please write for information.

WIDGEON COVE STUDIOS
c/o Georgeann Kuhl
RR1 Box 62
Harpswell, ME 04079
Tel: (207) 833 6081
Range of workshops in papermaking. Write for further details.

WOMEN'S STUDIO WORKSHOP
PO Box 489
Rosendale, NY 12472
Tel: (914) 658 9133
Workshops in papermaking and book arts. Write for further information.

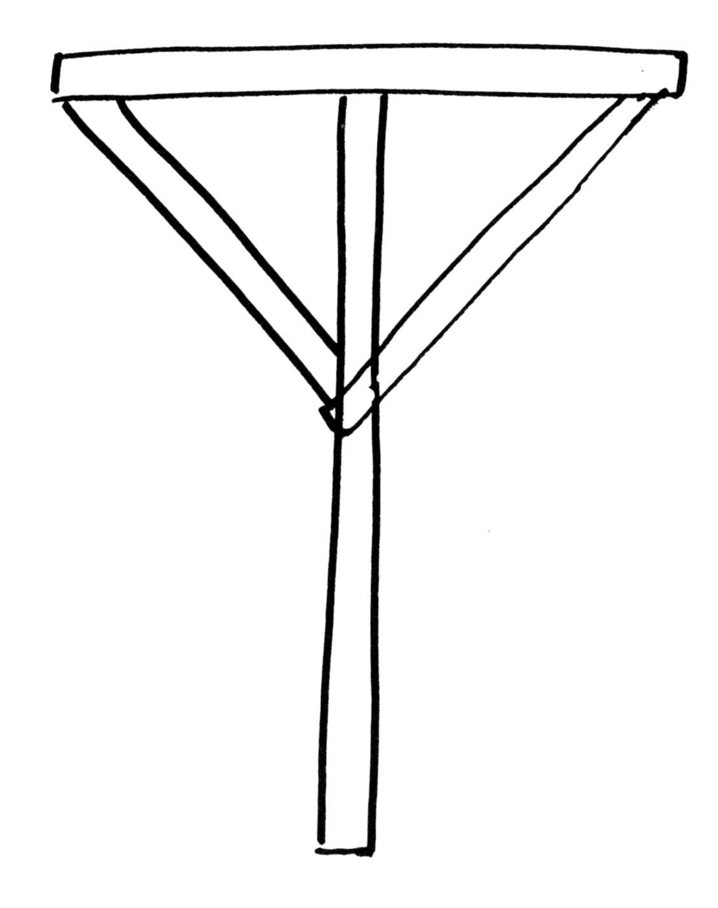

The variations of hand and mind permit endless permutations of quality because of the pliable nature of (paper's) simple ingredients and methods.

G. A. Beale *A Survey of Hand-made and Mouldmade Papers* Cadenza Press London 1977. (Out of print.)

Information resources

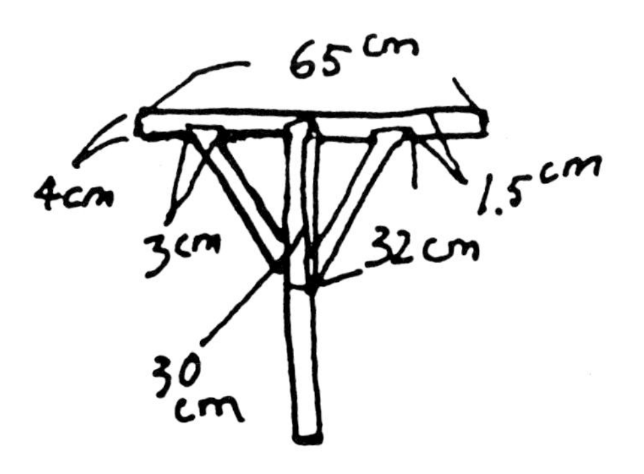

Information resources

To the numerous previous listings in this book, we have added groups and organisations currently offering lectures, advice and information about all forms of involvement with hand papermaking. Language restrictions have formed a partial barrier and this listing offers groups and organisations who operate essentially in the English language.

AUSTRALIA

FIBRES AND FABRICS GROUP
PO Box 329
Hermit Park, Qld 4812

GERALDTON PAPERMAKERS CLUB (GPC)
Bev Wasley (Hon Secretary)
PO Box 145
Geraldton, WA 6530
Formed in 1993, GPC members produce a variety of handmade papers and meet regularly to exchange ideas, information etc. Organises workshops both for members and non-members and, at various times throughout the year, hold displays and demonstrations, mainly at craft fairs.

PAPERMAKERS GUILD OF WESTERN AUSTRALIA
Gladys Dove
37 Denny Way
Alfred Cove, WA 6154
Tel: (09) 330 4102

PAPERMAKERS OF AUSTRALIA
(Publ. of *Words on Paper*)
Penny Carey-Wells
Papermill, Tasmanian School of Art
University of Tasmania, GPO Box 252C
Hobart, Tasmania 7001

PAPERMAKERS OF NEW SOUTH WALES
Jean Kropper
PO Box 1292, Lane Cove
Sidney, NSW 2066

PAPERMAKERS OF VICTORIA
Valda Quick (Secretary)
14 Dalmor Avenue
Mitcham, Vic 3132
Tel: (03) 523 0170

PRIMROSE PAPERWORKS CO-OPERATIVE LTD.
G. Berkemeir/J. Rubensohn (Directors)
PO Box 152
Cremorne, NSW 2090
Tel: (02) 909 1277
Based in Sidney, Primrose Paperworks runs activities in paper arts, calligraphy, bookbinding, printmaking,photography.

BELGIUM

LA MUSÉE NATIONAL DU PAPIER
Maisons Cavens
Place de Rome 11
B-4960 Malmedy
Tel: (2) 80 337058
The National Museum of Paper in Belgium has sections on history of papermaking, demonstrations of making paper by hand, and artifacts and machinery relating to papermaking. Other exhibits present the evolution of the Belgian paper industry from the time of the industrial revolution.

CANADA

ATELIER PAPYRUS INC.
523 Rue St. Paul
Trois Rivières
Quebec G9A 1H7
Tel: (819) 37 -0814
The Atelier Papyrus was established in 1984 as a non-profitmaking organisation dedicated to the promotion of hand papermaking and runs workshops, events, conferences and exhibitions and produces a range of speciality handmade papers together with some supplies for hand papermaking.

KAKALI HANDMADE PAPERS, INC.
1249 Cartwright Street
Vancouver, B.C. V6H 3R7
Tel: (604) 682 5274
This organisation functions principally as a paper institute and is the only one of its kind in Canada.

THE JAPANESE PAPER PLACE
887 Queen Street W.
Toronto M6J 1G5
Tel: (416) 369 0089
Fax: (416) 369 0163
Contact Nancy Jacobi. It is listed here to remind readers that over the world, there are paper shops that are vital centres of information and expertise.

REPUBLIC OF CHINA

SU HO MEMORIAL PAPER MUSEUM
Taipei, Taiwan
Ruey-Huey Chen, Director.

FINLAND

SUOMEN PAPERITAITEEN KESKUS
Finnish Paper Art Center Association
Kokonniementie 6
06100 Porvoo
Contact Väiski Putkonen, Chairman.
This organisation organises various paper-centred activities.

COSTA RICA

CENTRO DE INVESTIGACIONES EN FIBRAS Y PAPEL
Lil Mena, Apartado 103-1002
San Jose, Costa Rica
The Centre for Fibres and Paper Research held its first symposium on handmade paper in 1991. It operates a papermill, documentation centre, a studio for book arts, oriental and decorative papermaking.

EGYPT

PAPYRUS INSTITUTE
3 Nile Avenue
Giza, Cairo
Tel: (2) 989 476
Founded by Dr. Hassan Ragab, who rediscovered how to make papyrus.
Contact the centre for information.

FRANCE

MOULIN A PAPIER DU VAL DE LAGA À RICHARD DE BAS
F-63600 Ambert
Puy de Dôme
A museum and paper centre at this old mill. Contact the mill for information.

GERMANY

PAPER MUSEUM DÜREN
Leopold Hoesch Museum
Hoeschplatz 1
D-5160 Düren
Tel; (24 21) 121 25 29/121 25 60
Small working hand paper mill and museum. The International Biennale of Paper Art is held here. The museum has a commitment to exhibitions of papermaking and book arts.

ISRAEL

UNCLE BOB LESLIE PAPER MILL
POB 164
Omer
Tel: (011 972) 74695598
Centre for papermaking and study.

ITALY

LA FONDAZIONE MUSEO DELLA CARTA
Valle dei Mulini
Amalfi
La Fondazione opened in 1993 in one of the oldest surviving mills in Amalfi, dating to the 14th century. Historic papermaking equipment on display. Historic collection of papers, books and photographs. The focus of the museum is on research, especially papermaking in Italy. Also an exhibition programme and papermaking demonstrations.

JAPAN

KOCHI HANDMADE WASHI ASSOCIATION
Mr. Y. Morita (Representative)
110 Asahi-cho
3-chome Kochi, Kochi Pref.
Tel: (088) 7-3941
For information about the National Washi Equipment Preservation Society including makers of bamboo ribs, miscanthus ribs, weaving thread, screen frames, metal fixtures and brushes.

MINO HANDMADE WASHI CO-OPERATIVE ASSOCIATION
S. Goto (Representative)
T. Mame (Secretary)
777 Maeno
Mino, Gifu Pref.
Tel: (0575) 330 1241
For information on makers of screen frames, metal fixtures and brushes.

FUKUI WASHI PRODUCTION CO-OPERATIVE ASSOCIATION
M. Ishikawa (Representative)
M. Takayasu (Secretary)
11-11 Ohtaki, Imadate-cho
Imadate-gun, Fukui Pref.
Tel: (0778) 43 0875
For information on metal frames, watermarks and screen frames.

MALAWI

THE PAPER MAKING EDUCATIONAL TRUST (PAMET)
PO Box 1015
Blantyre, Malawi
Tel/Fax: 62 38 95
This trust was founded in 1991 with the aim of promoting and supporting small scale paper recycling in order to facilitate education and reduce poverty in Malawi.

NETHERLANDS

INTERNATIONAL ASSOCIATION OF HAND PAPERMAKERS AND PAPER ARTISTS (IAPMA)
Therese Weber (President 1994-96)
Sarah Comfort (Secretary 1994-96)
President Kennedylaan 150
Oegstgeest 2353 GV
IAPMA was founded in 1986 and is dedicated to assisting all those interested in paper. The aims and objectives of the association are to promote paper art and papermaking through the organisation of exhibitions and exchange programmes; to provide a resource centre comprising a slide collection of member's work; to stimulate contact and interaction between paper artists and papermakers by providing a meeting every two years, publishing a semi-annual Bulletin and a quarterly Newsletter. Subscription on an annual basis. Annual membership list. Write to the secretary for further information.

PHILIPPINES

DUNTOG FOUNDATION
Michael Parsons
PO Box 264
Baguio City
Tel/Fax: (74) 442 5555

FIBER DEVELOPMENT AUTHORITY
Philfinance Building
Benavidez Street
Legaspi Village
Makati, Metro Manila

SPAIN

MUSEU MOLI PAPERER DE CAPELLADES
E-08786 Capellades (Barcelona)
Tel/Fax: (3) 801 28 50
Inaugurated in 1961, this mill has, from its beginning, been a 'living' museum, with the working mill preserved in the basement (using the original tools and machinery). History of paper in Catalonia plus historical exhibitions demonstrating hand production in other countries. Also on display are papers from 13th-20th centuries.

SWEDEN

LESSEBO HANDPAPPERSBRUK
Lessebo Bruck
S-360 50 Lessebo
Lessebo papermill, dating back to 1693, still functions as a handmade paper mill. Papermaking courses for artists and handpapermaking are available.

SWITZERLAND

BASLE PAPERMILL
Swiss Paper Museum and Museum for Script and Printing
St. Alban-Tal 35/37
CH-4052 Basle
Tel: (61) 272 96 52
Fax: (61) 272 09 93
Installed in 1980, this is a working museum with a small papermill which includes the oldest preserved press dating from the fifteenth century. Further floors are devoted to the history of paper, writing, typesetting and bookbinding.

UK

BARCHAM GREEN & COMPANY LTD.
Hayle Mill
Maidstone, Kent ME15 6XQ
Tel: (01622) 692266
Fax: (0162) 756381
The Green family made paper for nine generations up until 1987 and Simon Barcham Green now offers commercial consultancy services including planning, feasibility studies, equipment specification and sourcing, layout, production techniques, productivity and quality control advice, costing and troubleshooting. The service is available worldwide on a fee plus expenses basis.

BRITISH ASSOCIATION OF PAPER HISTORIANS
c/o Dr John Griffiths
The Science Museum
Exhibition Road
London SW7 2DD
This organisation publishes a paper journal titled *The Quarterly*. For inquiries about 'The Quarterly' please contact the editor, Peter Bower at 64 Nutbrook Street, London SE15 4LE.

INSTITUTE OF PAPER CONSERVATION
Leigh Lodge
Leigh, Worcester WR6 5LB
Tel: (01886) 832218
Fax: (01886) 833688
The Institute of Paper Conservation is the leading body of professionals who conserve archives, books and works of art on paper. Although based in the UK, IPC has a worldwide membership which is also open to anyone interested in paper conservation. Details of members who can advise on paper conservation issues can be provided worldwide. Two international reply coupons would be appreciated with queries.

THE INTERNATIONAL PAPER HISTORIANS (IPH)
Dr Peter F. Tschudin (President)
Mrs Jenny Hudson (Editor)
4 Goldsmith Road
Walsall, West Midlands WS3 1DN
Tel: (01922) 496907
This is an international, non-profitmaking, scientific society whose aims include encouraging all efforts which are useful to paper historical research; promoting contacts and exchange of information on an international, national and regional basis; to improve the reputation of research and teaching and to encourage young academics to study paper history. Publishes a paper-historical bulletin and yearbook and organises scientific conferences and meetings of specialists. Subscription on an annual basis.

PULP AND PAPER INFORMATION CENTRE
Papermakers House
Rivenhall Road
Swindon SN5 7BE
Tel: (01793) 886086
Fax: (01793) 886182

WOOKEY HOLE PAPER MUSEUM
Wookey Hole
Wells, Somerset
Tel: (01749) 672243
Old papermaking mill which is still operational, with paper museum.

USA

PAPER SOURCE
232 West Chicago Ave
Chicago, Ill 60610
Tel: (312)337 0798
also at
1810 Massachusetts Ave
Cambridge MA 02140
Tel: (617) 497 1077
One of the largest retail sources of handmade papers, carrying sheets from almost every country. Stores regularly sponsor on-site workshops.

THE RESEARCH INSTITUTE OF PAPER HISTORY & TECHNOLOGY
Carriage House
8 Evans Road
Brookline, MA 02146
Tel: (617) 232 1636
Fax: (617) 277 7719
Contact Elaine Koretsky. This new research facility was founded on October 1 1994 by Elaine and Donna Koretsky. It houses a complete hand papermaking facility; a library of old, rare and modern books dealing with paper history, the technology of papermaking and manuscripts illustrating book forms in many old cultures; and a collection of tools, equipment, handmade paper and artifacts relating to historical papermaking which have been gathered from all over the world.

FRIENDS OF DARD HUNTER
Box 50, HCR 34
Montpelier, VT 05602
Dard Hunter (1883-1966) was an artist, designer, craftsman, papermaker, traveller, collector, writer and historian whose writings and museum collection on papermaking and printing provide inspiration for the Friends of Dard Hunter. "A unique group of enthusiasts from hand papermakers to paper historians to book artists to paper scientists to paper marblers to archivists." Publishes a periodic newsletter and an annual Membership Directory. Annual meeting for members to exchange information.

Now having mechanised your mill
the ardent craftsmen must be found . . .
Such men who love and live their work
will jump from bed to toil like gods
in the dark cumbersome shadows of the night
undeterred by finger-tip frost or
sweltering midsummer limbs.
Such Titans don't succumb to sleep
when paper must be pounded out.

17th century poem, *Papyrus* by Father Imberdis

Glossary of papermaking terms

Left Robbin Amy Silverberg, Dobbin Mill, New York
Middle Dieu Donné Papermill Inc., New York
Right Jacki Parry, The Paper Workshop, Glasgow

'A' SERIES — International ISO range of paper sizes reducing from AO (841 x 1189 mm) by folding in half to preserve the same proportions of 1:√2 at each reduction.

ABSORBENCY — The degree in which paper takes up contact moisture measured by a standard test.

AIR DRIED — Dried with hot or cold air; it can include loft drying or machine drying.

ALL-RAG PAPER — Paper made from rags as the basis for the pulp. Today can mean cotton linter pulp.

ALUM — A complex salt, most commonly aluminium sulphate added with rosin to the pulp while it is in the beater as a sizing agent to impart a harder and more water-resistant surface to the finished sheet; also acts as a preservative and a mordant for fixing colours. If not removed from the fibre, its high acidity will cause irreversible damage to the paper.

'B' SERIES — International ISO range of sizes designed for large items and falling between A series sizes.

BEATING — Hand or mechanical maceration of fibres to prepare them for pulp. Part of papermaking process where fibres are mechanically treated to modify their characteristics to those required by desired paper quality in manufacture.

BRIGHTNESS — The paper's ability to reflect white light.

BUFFERING AGENT — Also termed 'alkaline reserve', it is an alkaline substance, usually calcium carbonate or magnesium carbonate, occurring naturally in a water supply or purposely added by the papermaker to help protect the paper from acidity in the environment.

CALENDERING — A pressing process used to smooth or glaze a sheet of paper during the finishing process. A set of rollers on a paper machine which give a smooth finish to the web as it passes through by applying pressure. Calendered paper has a smooth, medium-gloss finish.

CELLULOSE — The main part of the cell wall of a plant.

CONTRARIES — Unwanted pieces of materials which have become embedded in a sheet of paper.

CUSTOM MAKING — A paper made specially for a client; opposite to a standard range.

DECKLE EDGES — The wavy, feathered or ragged edges of a sheet of handmade paper on four sides caused by the deckle frame where pulp thins towards the edge of the mould. Also found on cylinder mouldmade papers on the two outside edges of the web.

DYES — Water-soluble colouring agents which usually penetrate and become attached to the fibre. Types of dyes include: direct dyes, organic dyes usually derived from coal tars; fibre-reactive dyes, that form a chemical bond with the fibres; and natural dyes, derived from natural sources such as indigo and onion skins. Since some types of dyes require an acid mordant to set or fix the dye to the fibre, care must be used in their selection.

FASTNESS — Resistance of colour to fading.

FELT	A rectangular sheet of absorbent material, usually of wool, cut in a larger size than the paper sheet required. It is utilised during the Western making process when newly formed sheets of paper are couched or transferred from the mould.
FELT FINISH	Paper may be naturally dried after pressing and may acquire the texture of the surface of the felt covering the paper.
FIBRES	The basis of a sheet of paper. Papermaking fibres are hollow tube-like structures with walls made up of 'fibrills'. (See PAPERMAKING FIBRES.)
FIBRILLATION	Shredding and bruising of fibre walls during beating process.
FILLER	A material generally added to the beater to fill in the pores of a fibre, making a harder, more opaque surface. Pigments added to the furnish of paper to improve the printing characteristics and appearance of the printed image.
FINISHING	Practices of drying, sizing and looking over sheets of paper after the making processes.
FORMATION	The fibre distribution in a sheet of paper as it appears when held up to the light.
FURNISH	The ingredients in the beater, which, when added together, give a specific type of paper.
GELATINE	A type of size obtained from animal tissues applied to the surface of paper to make it impervious to water and to aid resistance to bleeding during printing. Also imparts surface strength to watercolour and drawing papers.
GLAZING	The term used to denote a smooth surface given to a sheet of paper often made by running dried sheets through steel rollers or between polished zinc plates.
GRAMMAGE (gsm/gm^2)	The weight of paper and board expressed in metric terms, grams per square metre.
HANDMADE	Paper made by hand operation.
HYDRATION	A process occurring during beating in which the bruised fibres begin to accept water.
KYOSEI-MAKING	"*Konnyaku* sized and crumpled paper then cooked in lime powder solution". From *Kami* '89 by Asao Shimura.
LAID PAPER	Paper that is made on a laid mould. It is customary for the laid lines to run across the page's width and the chain (linking) lines from head to foot.
MATURE	To allow paper to settle after making.
MOULD	The basic tool of a Western hand papermaker which consists of a (mahogany) frame and removable deckle. Varying designs exist in different countries.
PULP	The 'stuff' used in papermaking process (in Japan, it always refers to wood pulp). Chemical pulp contains many fewer impurities than mechanical pulp.

PURITY	The degree of chemical purity of a paper.
QUIRE	A quantity of 25 sheets of paper, a twentieth part of a ream.
REAM	Traditionally 480 sheets (equal to 20 quires of 24 sheets). Now taken to refer to 500 sheets.
RELATIVE HUMIDTY (RH)	Amount of water vapour present in the atmosphere expressed as a percentage of saturation. The moisture content of paper is governed by the relative humidity of the surrounding atmosphere. Research (and experience) has shown that paper perhaps has its best dimensional stability in the range of 40 - 55% relative humidity.
ROSIN	An internal sizing agent for paper. Occasionally used as a tub size. It is acidic in nature and detrimental to the permanence of paper.
ROUGH	A term used to describe the surface texture of a sheet of handmade paper.
SIZING	A solution used to make paper moisture-resistant in varying degrees. Size can be added at two stages of the papermaking process (see below). The degree of sizing of paper determines its resistance to the penetration of moisture. Internal Sizing, Engine-Sizing, Beater-Sizing describes moisture-resistant pulps which receive sizing treatment in the beater. Tub Sizing - after manufacture, when the paper has been dried, some papers are passed through a solution of gelatine (or other size) traditionally contained in a bath or tub. Other surface sizes include glue, casein and starch. Animal Tub Sizing - Refers to the papers that are tub sized and for which the gelatine used is obtained from a solution of parings and skins of animals.
SPECIAL FURNISH	Papers made from a special mixture of pulps for a specific purpose.
STOCK	Suspension of fibres in water.
VATMAN	A person engaged in the scooping of pulp from the vat to form a sheet of paper.
WATERLEAF	A term used to describe paper that contains no sizing, and therefore is generally absorbent.

There is always more to learn on the papermaking road . .

Nigel Macfarlane *A Paper Journey* 19

List of suppliers by country

ARGENTINA
MOLINO DEL MANZANO, EL
GEILER, ALEJANDRO D.

AUSTRALIA
GERALDTON PAPERMAKERS
NICHOLLS, FRED
OYSTER BAY PAPER CRAFTS
PAPER CAPERS
PAPERMAKERS GUILD OF WESTERN AUSTRALIA
PAPERMAKERS OF AUSTRALIA
PAPERMAKERS OF NEW SOUTH WALES
PAPERMAKERS OF VICTORIA
PAPER MERCHANT, THE
PRIMROSE PAPERWORKS
PROSSER, LOIS & BARRY
STEAM FILM PTY LTD.

BELGIUM
DECOSTER, JEAN
LA MUSÉE NATIONAL DU PAPIER

CANADA
ATELIER PAPYRUS INC.
CRANBERRY MILLS
FLAX PAPER WORKS
HEMPLINE INC.
JAPANESE PAPER PLACE, THE
KAKALI HANDMADE PAPERS INC.
MIKOLET STUDIOS
NEW LEAF PAPERMILL
PAPERTRAIL, THE
PAPETERIE SAINT-ARMAND, LA

REPUBLIC OF CHINA
SU HO MEMORIAL PAPER MUSEUM

COSTA RICA
CENTRO DE INVESTIGACIONES EN FIBRAS Y PAPEL

DENMARK
AV-FORM
OPPENHEJM, HELMUTH
VILSBØLL, ANNE

EGYPT
ASSOCIATION FOR THE PROTECTION OF THE ENVIRONMENT
PAPYRUS INSTITUTE
ZABBELEEN PAPER CENTRE

FINLAND
SUOMEN PAPERITAITEEN KESKUS

FRANCE
RICHARD DE BAS, MOULIN Å PAPIER DU VAL DE LAGA

GERMANY
DRUCKERN & LERNEN GMBH
EIFELTOR MÜHLE
FRERICK, HELMUT
GLASER, ANTON
PAPER MUSEUM, DÜREN , (LEOPOLD HOESCH MUSEUM) P
APIERWERKSTATT JOHN GERARD
PAPIERWERKSTATT ALFRED KÖNIG
TEMMING, PETER AG

HAWAII
WOLD, MARILYN

INDIA
BRAMCO ENGINEERS
KHADI PAPERS INDIA
PORRITT AND SPENCER (ASIA) LTD.
SITSON INDIA
SRI AUROBINDO ASHRAM
UNIVERSAL ENGINEERING CORPORATION

ISRAEL
UNCLE BOB LESLIE PAPER MILL

ITALY
FONDAZIONE MUSEO DELLA CARTA, LA

JAPAN
ARIMITSU, HR.
DIAFLOC CO.
FUJI PAPER MILLS CO-OPERATIVE
FUKUI WASHI PRODUCTION CO-OPERATIVE ASSOCIATION
KIKUCHI, HR.
KINUYAMA, YOSHIROU
KOCHI HANDMADE WASHI ASSOCIATION
MINO HANDMADE WASHI CO-OPERATIVE ASSOCIATION

MALAWI
PAPERMAKING EDUCATIONAL TRUST, THE (PAMET)

NETHERLANDS
BUCHEL VAN DER KORPUT B.V.
GENTENAAR, PETER
INTERNATIONAL ASSOCIATION OF HAND PAPERMAKERS AND PAPER ARTISTS (IAPMA)

NEW ZEALAND
VICTORIA HOUSE PAPER

PHILIPPINES
BERNARD HANDMADE PAPER PRODUCTS
DUNTOG FOUNDATION
FIBRE DEVELOPMENT AUTHORITY
JARNET TRADE INTERNATIONAL
KAMI PAPER MAKERS, PHILIPPINES

SOUTH AFRICA
PAPIER DU PORT, LE
TECHNIKON NATAL

SPAIN
CAPELLADES, MUSEU MOLI PAPERER DE
CELESA
CELLULOSA DE LEVANTE SA
ENCE
MONTFORT DELMAS, JULIO, SA
TALLERAS SOTERAS

SWEDEN
HOLM, SANNY
KLIPPANS FINPAPPERSBRUK, AB
LESSEBO HANDPAPPERSBRUK
LORENTZE & WETTRE
ZEN ART PAPER AB

SWITZERLAND

BASLE PAPER MILL
PAPIERATELIER
SANDOZ PRODUCTS (SCHWEIZ) AG
FONTAINE, VIVAINE
WEBER, THERESE

UK

AMIES, EDWIN & CO LTD.
ATLANTIS EUROPEAN LTD.
BARCHAM GREEN & COMPANY LIMITED
BRITISH ASSOCIATION OF PAPER HISTORIANS, THE
BRITISH PAPER COMPANY, THE
CONSERVATION RESOURCES (UK) LTD.
CRAFTS
DAWSON, SOPHIE
FALKINER FINE PAPERS
GLASGOW SCHOOL OF ART
HERCULES LTD.
HOPE EDUCATION LTD.
INSTITUTE OF PAPER CONSERVATION
INTERNATIONAL PAPER HISTORIANS, THE
JOHN PURCELL PAPER
KHADI PAPERS
McMILLAN, T.
MERCK LTD.
MESSMER INSTRUMENTS LTD.
MONKMAN, FRANK LTD.
NAUTILUS PRESS & PAPER MILL
P&S TEXTILES
PAPER PLUS
PAPER WORKSHOP, THE
PLANT PAPERS
PRESERVATION EQUIPMENT LTD.
PULP AND PAPER INFORMATION CENTRE
SANDOZ CHEMICALS (UK) LTD.
SPECIALIST CRAFTS LTD.
SPIRES, GILLIAN HANDMADE PAPER
TEST PAPERS DIVISION
TIRANTI
WOOKEY HOLE PAPER MUSEUM

USA

AMERICAN CRAFTS
ARIZONA WORKSHOPS
BARRETT, TIMOTHY
CARRIAGE HOUSE PAPER
CENTRE FOR BOOK ARTS
CENTRE FOR SAFETY IN THE ARTS
COLOPHON BOOK ARTS SUPPLY
COLUMBIA COLLEGE, CHICAGO
CONSERVATION RESOURCES INTERNATIONAL, INC.
CREATIVE ARTS WORKSHOP
DARD HUNTER, FRIENDS OF
DIEU DONNÉ PAPERMILL, INC.
DOBBIN MILL
FABRILE STUDIO
FARRIN & FLETCHER DESIGN STUDIOS
FIBREARTS
FIBRES AND FABRICS GROUP
GOLD'S ARTWORKS, INC.
GORMLEY, JAMES F.
HAND PAPERMAKING
HAND PAPERMAKING SLIDE KITS AND VIDEO
HISTORIC RITTENHOUSETOWN
ICOSA STUDIO
INTER-OCEAN CURIOSITY STUDIO
LEN KEL MANUFACTURING
McDONALD, LEE S., INC.
MAGNOLIA EDITIONS
McGOWAN FILM & VIDEO
MARALEX STUDIOS
MARKIM, GREG INC.
MINNESOTA CENTER FOR BOOK ARTS
MOORE, TIMOTHY (See PARAGON PAPER MOLDS)
PAPER ARTS MILL
PAPER CRAFTERS
PAPER SOURCE
PAPERMAKING VIDEOTAPES
PARAGON PAPER MOLDS
PRO CHEMICAL AND DYE, INC.
PULPERS, THE
PYRAMID ATLANTIC
REINA, DAVID, DESIGNS INC.
RESEARCH INSTITUTE OF PAPER HISTORY AND TECHNOLOGY, THE
RUGG ROAD PAPERS AND PRINTS
SCHECK, BARBARA
SEA PENN PRESS AND PAPER MILL
STRAW INTO GOLD
SUBMARINE PAPERWORKS
TALAS
TAOS PAPERWORKS
TOCKINGTON PAPER STUDIO
TWINROCKER HANDMADE PAPER
UNIVERSITY OF IOWA
UNIVERSITY PRODUCTS LTD.
VISUAL STUDIES WORKSHOP
WATERLEAF MILL & BINDERY
WIDGEON COVE STUDIOS
WOLD, MARILYN
WOMEN'S STUDIO WORKSHOP

Suppliers index

Photographic and illustration acknowledgements

P2, Drawing by Steven Drewett; P4 Photo courtesy Kathryn Clark; P7 Drawing by Steven Drewett; P9 Drawings courtesy of Asao Shimura; P9 Photo courtesy Sophie Dawson; P13 Photos courtesy Helmut Becker and Silvie Turner; P16 Drawing by Steven Drewett; P33 Photos courtesy Anne Vilsbøll and David Reina Designs; P34 Photo credit Silvie Turner; P 36 Photo courtesy Peter Gentenaar; P38 Photo courtesy Lee S. Mcdonald; P40 Drawings courtesy Maureen Richardson; P42 Photo credit Silvie Turner; P45 Drawing by Steven Drewett; P49 Photos courtesy Claudia Stroh Gerard, Nigel Macfarlane, photo credit Silvie Turner; P50 Photo courtesy Peter Zerbe, photo credit Silvie Turner; P61 Photo courtesy Simon Green; P63 Photos courtesy Nigel Macfarlane, Maureen Richardson; P64 Photo courtesy Oyster Bay Paper Crafts; P69 Drawings courtesy of Asao Shimura; P70, 71, 73 Photo credits Silvie Turner; P76 Drawing by Steven Drewett; P78 Photo courtesy Nigel Macfarlane; P83 Photo courtesy Helmut Frerick; P88 Drawings by Steven Drewett (inspired by Jules Heller); P95 Photo courtesy John Gerard; P101 Drawings courtesy Asao Shimura; P105 Photos courtesy Robbin Ami Silverberg, Dieu Donné Press and Papermill and Jacki Parry, photo credit Robert Connelly.